Electronics Workshop Companion for Hobbyists

Stan Gibilisco

New York Chicago San Francisco
Athens London Madrid
Mexico City Milan New Delhi
Singapore Sydney Toronto

McGraw-Hill Education books are available at special quantity discounts to use as premiums and sales promotions or for use in corporate training programs. To contact a representative, please visit the Contact Us page at www.mhprofessional.com.

Electronics Workshop Companion for Hobbyists

1 2 3 4 5 6 7 8 9 0 DOC/DOC 1 2 0 9 8 7 6 5

ISBN 978-0-07-184380-5
MHID 0-07-184380-9

This book is printed on acid-free paper.

Sponsoring Editor
Roger Stewart

Editorial Supervisor
Donna M. Martone

Production Supervisor
Pamela A. Pelton

Acquisitions Coordinator
Amy Stonebraker

Project Manager
Nancy Dimitry

Copy Editor
Nancy Dimitry

Proofreader
Don Dimitry

Art Director, Cover
Jeff Weeks

Composition
Gabriella Kadar

Indexer
WordCo Indexing Services

In memory of Jack,
mentor and friend

About the Author

Stan Gibilisco, an electronics engineer and mathematician, has authored multiple titles for the McGraw-Hill *Demystified* and *Know-It-All* series, along with numerous other technical books and dozens of magazine articles. His work appears in several languages. Stan has been an active Amateur Radio operator since 1966, and operates from his station W1GV in the Black Hills of South Dakota, USA.

Contents

Introduction

If you like to invent, design, build, test, and tweak electronic circuits and gadgets, then you'll like this book. It's aimed at beginning and intermediate-level hobbyists and home experimenters, although technicians and engineers should find it useful as a reference from time to time.

Chapter 1 offers suggestions for setting up a basic home electronics workshop: a sturdy bench, plenty of organized storage space, a test meter, an ample supply of components and connectors, a modest computer, and a reliable source of electricity. Once you have your lab together, I'll show you a simple experiment you can do there.

Chapter 2 describes the types of resistors available for use with electronic circuits, defines some of the more technical jargon that you'll encounter, offers a few useful formulas for resistance calculations, and concludes with three simple experiments you can do without spending a lot of time or money.

Chapter 3 offers a brief refresher on capacitance, along with information about the types of capacitors you can easily obtain and use in your electronics adventures. You'll learn a few capacitance-relevant formulas. Two experiments will give you some insight into how capacitors behave and how to measure their values.

Chapter 4 is the inductance counterpart to Chapter 3. You'll learn about coil core types and optimum coil configurations for audio versus radio-frequency circuits. You will also be shown some simple inductance calculations. In the experiments, you'll build a simple DC electromagnet and an inductor-based galvanometer.

Chapter 5 involves transformers and their uses, from changing voltages to matching impedances. You'll learn formulas to help you choose the best transformer for your evolving creation. You'll test a small transformer module and then conduct an experiment where you connect two identical modules "back-to-back."

Chapter 6 gives you an overview of diode applications including rectification, frequency multiplication, signal mixing, switching, voltage regulation, amplitude limiting, frequency control, oscillation, and DC power generation. You'll do a couple of experiments to reduce DC voltage and convert AC to DC.

Chapter 7 deals with the fundamentals of bipolar and field-effect transistors, including metal-oxide devices. You'll learn which transistors work best in particular circuits. In the experiments, you'll use a multimeter to test a bipolar transistor and a junction field-effect transistor (JFET) for proper operation.

Chapter 8 describes integrated circuits (ICs), emphasizing the advantages but noting the limitations. You'll get familiar with linear versus digital IC technology, learn IC functions, and fortify your understanding of binary logic hardware. In the experiments, you'll use resistors and diodes to simulate the operation of OR and AND gates.

Chapter 9 offers an assortment of hardware manipulation and lab techniques, such as cells and batteries, wire and cable splicing, soldering and desoldering, commonly used connectors, oscilloscopes, spectrum analyzers, frequency counters, and signal generators.

An extensive set of Appendixes contains diverse reference data involving electrical and electronic hardware, and breaks down the radio-frequency (RF) spectrum into its formally defined bands.

With this reference in your workshop library, you'll have the fundamental information needed to undertake your odyssey into the world of hobby electronics, from hi-fi to ham radio, from switches to microprocessors.

I welcome your suggestions for future editions. Please visit my website at **www.sciencewriter.net**. You can e-mail me from there.

Above all, have fun!

Stan Gibilisco

CHAPTER 1
Setting Up Shop

As an electronics hobbyist, you'll want a site that can stand up to plenty of activity. If you own a home with a basement, then you have an ideal place for a workshop waiting for you to add some imagination and "sweat equity." If you live in a condo, an apartment, or a home without a basement, you'll have more trouble setting up your shop, especially if you share limited living space. Nevertheless, true electronics enthusiasts never fail to find a place to carry on their art.

Workbench

Before a writing and video-production computer claimed it, my electronics workbench comprised a piece of plywood, weighted down over the keyboard of an old upright piano in the cellar, and reinforced by chains from the ceiling. As I sit on a barstool four feet above the floor to write this book, I gaze over the top of a computer tower to see my new, less elaborate electronics workstand on the top shelf of a general-purpose, heavy-duty storage set that I bought at a local department store.

Locating It

Your test bench doesn't have to be as unorthodox as mine, of course, and you can put it anywhere you want, as long as it won't shake or collapse. The surface should consist of an electrically nonconductive material, such as wood or hard plastic, protected in the work area by a meat-cutting board and/or baking sheet, as shown in Fig. 1-1. You'll want some containers for electronic components, and a place to plug in electrical tools, such as the soldering iron shown in the figure. A desk lamp with an adjustable arm completes the ensemble; mine is out of sight here, affixed to the wall.

FIGURE 1-1 My scaled-down electronics workbench occupies the top level of a general-purpose shelf set.

> **Tip**
>
> Buy a pair of safety glasses at your local hardware store, preferably the sort that welders use. Wear them at all times while building, experimenting, testing, and troubleshooting. You never know when a wire fragment will fly at your eyes when you snip it off, or heated solder will spatter more than you thought it could, or you'll inadvertently touch two live wires together and get a spark that travels farther than you imagined possible! On several occasions, I've had my safety glasses save me trips to the hospital emergency room.

Storage

Somewhere near your workbench, you should have some small storage bins with numerous drawers for small components, such as resistors, capacitors, diodes, transistors, and the like. I have three of these cabinets, measuring about 12 inches wide, 15 inches high, and 6 inches deep. Figure 1-2 shows my arrangement, along with speakers connected to my ham radio station's enhanced sound system, on a storage shelf across the room from my main workbench. Figure 1-3 is a close-up view of a couple of open drawers in one of these bins.

General Supplies

Table 1-1 lists some items that you'll most likely want for a beginner's hobby electronics workshop. You can find most of these components at hardware and department stores. In a few cases, you might need to visit a Radio Shack store or order a component from their website at www.radioshack.com.

Figure 1-2 Storage bins for small electronic components and hardware items.

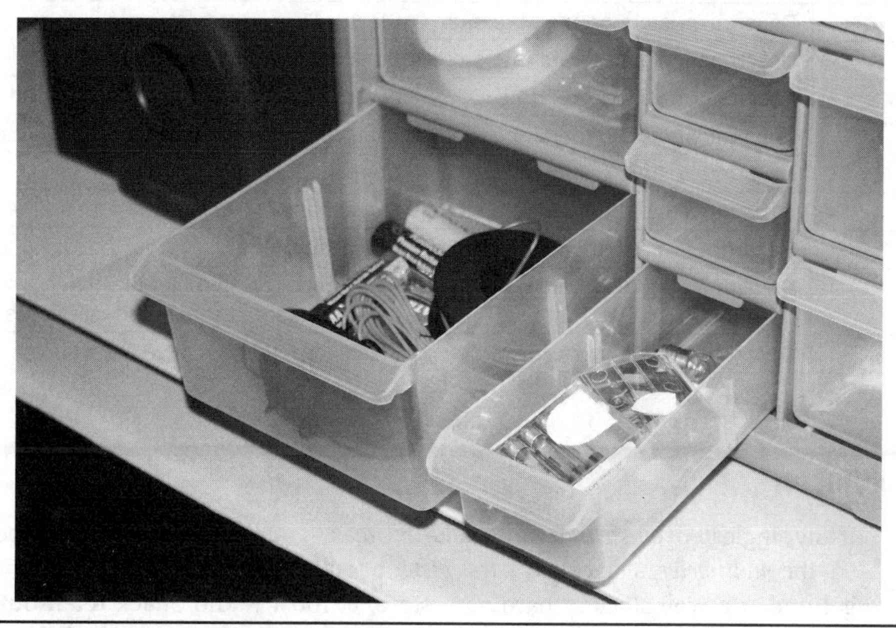

Figure 1-3 Drawers in storage bins allow for easy access and concealment.

TABLE 1-1 Suggested Basic Items for a Beginner's Electronics Workshop

Quantity	Source	Description
1	Department store or hardware store	Pair of safety glasses that seal all the way around your eyes
1	Grocery store or department store	Large wooden meat-cutting board or aluminum cookie sheet to protect work surface
1	Department store or hardware store	Basic tool kit (hammer, screwdrivers, wrenches, etc.)
1	Department store or hardware store	Large tool box
1	Department store	Pair of rubber-soled shoes
1	Department store	Pair of rubber gloves
1	Department store	12-inch plastic or wooden ruler
1	Department store or hardware store	36-inch measuring stick
1	Department store or hardware store	Large magnifying glass with handle
1	Hardware store or Radio Shack	Electrical test meter (volt-ohm-milliammeter) with digital numeric readout
1	Hardware store or Radio Shack	Electrical test meter with analog readout
1	Hardware store or Radio Shack	Soldering gun rated at about 150 watts
1	Hardware store or Radio Shack	Soldering iron rated at about 50 watts
1	Hardware store or Radio Shack	Roll of rosin-core solder for electronic circuits, 40% tin and 60% lead
3	Department store or hardware store	Large roll of electrical tape
1	Department store or hardware store	Diagonal wire cutter/stripper
1	Department store or hardware store	Small needle-nose pliers
1	Radio Shack Part No. 278-1156	Package of insulated test/jumper leads
1	Department store or hardware store	Multi-outlet cord with transient suppressor
5	Department store or hardware store	Steel wool pad without soap
5	Department store or hardware store	Emery paper or emery cloth
1	Department store or hardware store	LED flashlight
3	Department store or hardware store	Multi-drawer cabinets for electronic components and small hardware items
4	Department store or grocery store	Plastic or glass 16-ounce measuring cups

Multimeter

If any single item exists that can "make or break" your hobby electronics experience, it's the *multimeter*, also called a *test meter* or *volt-ohm-milliammeter* (VOM). You can get one at a well-stocked hardware store, at most Radio Shack retail outlets, or through various websites. These meters come in two main types. *Analog* meters

Figure 1-4 At left, an analog multimeter. At right, a digital multimeter.

have needle-and-scale readouts. *Digital* meters show you the numbers. The type that you buy depends on your personal preference. I have both an analog multimeter (at the left in Fig. 1-4) and a digital multimeter (at the right in Fig. 1-4).

Warning! **Whenever you use a multimeter to test a circuit where the voltage will exceed 12 V (the level produced by an automotive battery), wear rubber gloves and a pair of shoes with electrically insulated soles. That way, you'll ensure that you can't receive a dangerous shock. In addition, by wearing gloves, you insulate yourself completely from the circuit under test, ensuring that your body's *internal resistance* can't throw off the meter reading. This exasperating phenomenon can occur when you measure high resistance values or tiny currents.**

Analog Meter

My analog multimeter has several graduated scales and a rotary switch with 14 settings. When the switch points straight up, the meter is turned off. Going clockwise from the OFF position, the switch allows measurement of the following quantities, in order:

1. DC volts (DCV) from 0 to 10
2. DCV from 0 to 50
3. DCV from 0 to 250
4. DCV from 0 to 500

5. AC volts (ACV) from 0 to 500
6. ACV from 0 to 250
7. ACV from 0 to 50
8. DC milliamperes (DCmA) from 0 to 250, where one milliampere (1 mA) equals a thousandth of an ampere (0.001 A)
9. DCmA from 0 to 25
10. Battery test (BAT) for 1.5-volt (1.5-V) cells
11. Battery test (BAT) for 9-volt (9-V) batteries
12. Resistance in thousands of ohms (×1k)
13. Resistance in tens of ohms (×10)

Tip

Keep your multimeter meter powered-down when you're not using it. That meter, even though it seems passive, contains a battery that will gradually discharge if you leave the meter powered up, even if it sits idly on a shelf.

Tip

When you measure resistance, the meter reading depends on the amount of current passing through the device that you're testing. The meter's internal battery forces a certain current through the component, and that current depends on the resistance. As the resistance goes down, the current goes up, so the meter's resistance scale goes backwards. The sideways 8 (∞) at the left-hand end of the scale means "infinity ohms." That's an open circuit, where no current flows at all.

Resistance Calibration

On an analog meter that can measure resistance, you'll find a little knob that allows you to adjust the meter for the correct zero reading. It will say 0Ω ADJ or something like that. The horseshoe symbol is an uppercase Greek letter *omega*, which stands for *ohms*. To use this control, set the meter switch to the resistance range that you intend to use, short the red and black test probes directly together, and turn the knob until the meter needle goes all the way to the right-hand end of the resistance scale and comes to rest at the hash mark for 0 ohms. Don't turn the knob past that point. The needle should hover exactly over the 0 marker. Use this adjustment control whenever you change the meter from one resistance scale to another, and also if you haven't used your meter for a while. As the battery grows weak, the meter's accuracy will degrade unless you "tweak" this control before every resistance test.

Tip

When you want to use a multimeter to measure current or voltage, start with the highest meter scale and work your way down. That way, if the quantity that you want to measure exceeds the maximum scale value, the meter's needle won't slam against the pin at the top end.

Warning! **When you want to measure the resistance between two points in a circuit, make sure that the device under test is switched off before you connect a multimeter to it. Otherwise, you'll probably get an inaccurate reading. You might even damage your meter, disrupt the operation of the circuit under test, or both.**

Digital Meter

My digital multimeter has a rotary switch with 20 positions. As with the analog meter, the top switch position represents OFF. Going clockwise from there, the switch allows for measurement of a variety of quantities, in this order:

1. AC volts (V~) from 0 to 500, with a special insert for the red test probe wire
2. V~ from 0 to 200
3. DC amperes (A—) from 0 to 200 microamperes (the switch says 200μ), where one microampere (1 μA) equals a millionth of an ampere (0.000001 A)
4. A— from 0 to 2000 μA (the switch says 2000μ)
5. A— from 0 to 20 mA (the switch says 20m)
6. A— from 0 to 200 mA (the switch says 200m)
7. A— from 0 to 10, with a special insert for the red test probe wire (the switch says 10A)
8. A blank spot that doesn't do anything as far as I know
9. Diode test (a diode should conduct in one direction but not the other)
10. DC resistance in ohms (Ω) from 0 to 200
11. Ohms (Ω) from 0 to 2000 (the meter says 2k)
12. Ohms (Ω) from 0 to 20,000 (the meter says 20k)
13. Ohms (Ω) from 0 to 200,000 (the meter says 200k)
14. Ohms (Ω) from 0 to 2,000,000 (the meter says 2000k)
15. DC voltage (V—) from 0 to 200 millivolts (the switch says 200m), where one millivolt (1 mV) equals a thousandth of a volt (0.001 V)
16. V— from 0 to 2000 mV (the switch says 2000m)
17. V— from 0 to 20 volts (the switch says 20)
18. V— from 0 to 200 volts (the switch says 200)
19. V— from 0 to 600 volts, with a special insert for the red test probe wire (the switch says 600)

Tip

The red test probe wire should always go to the more positive point in a DC circuit or system, whether you measure current or voltage. In an AC situation, it doesn't matter which probe goes where.

Tip

If you really want to know how to get the best use out of your multimeter, read the instruction manual that came with it. You'll learn something new every time you look at that manual. You might also avoid a mistake that could damage something.

Bonus Equipment

Table 1-1 shows only the most essential items that you'll need to get started with hobby electronics experimentation. As you gain experience and knowledge, you can shop for more tools and equipment.

Breadboard

In the early years of electronics, experimenters used circuit boards fabricated from wooden slabs meant for bread dough preparation. They called them *breadboards*, and the term has endured to this day. I cobbled a wooden board with nails pounded into it for activities in my book, *Electricity Experiments You Can Do at Home* (McGraw-Hill, 2010). Figure 1-5 is a pictorial diagram of the layout that I concocted.

Some people laugh at the breadboard that I made from wood. Indeed, the contraption looks ancient next to integrated circuits, microcomputers, and other latter-day electronic components. With many of today's ultraminiature components, the thing is too large, although it works okay for heavy-duty items, such as toroidal coils, variable capacitors, and power-supply transformers.

If you'd rather go small, you can select from various perforated boards at Radio Shack stores, or you can use one that comes with a kit, such as the *Vilros* ensemble for the *Arduino* programmable microcomputer (Fig. 1-6). But you might find such a breadboard so compact that it gives you trouble. I need a magnifying glass to see the parts that I place in the microscopic holes on the Vilros board, and a steady hand to guide the components and connectors into the correct holes.

Computer

You can do some electronics experiments without a computer, but if you're like most hobbyists, you'll want one in your shop. You can easily move a small notebook

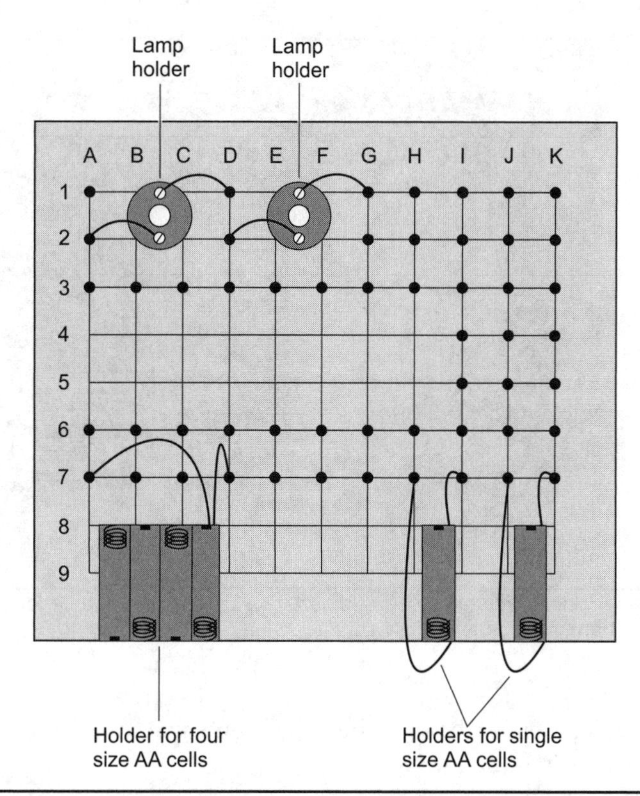

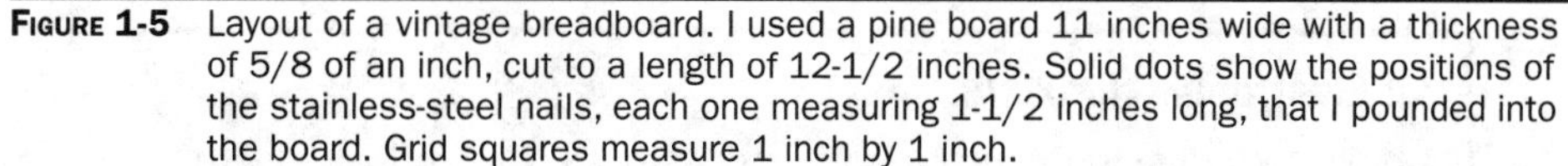
FIGURE 1-5 Layout of a vintage breadboard. I used a pine board 11 inches wide with a thickness of 5/8 of an inch, cut to a length of 12-1/2 inches. Solid dots show the positions of the stainless-steel nails, each one measuring 1-1/2 inches long, that I pounded into the board. Grid squares measure 1 inch by 1 inch.

computer around, and if you have a home wireless network, you can use it to get information from the Internet. For example, you might want to locate a source for an exotic component, such as an air-variable capacitor. If you have a computer handy, you can "google" a term or phrase that closely describes what you're looking for, and you'll usually come up with supplier options in a matter of seconds. I got an air-variable capacitor from a source called *Amplified Parts* (www.amplifiedparts.com) as a result of that sort of search. That vendor, by the way, is a great source of parts for hobby electronics in general.

For hobby work involving electronics experiments and simulations with programmable externals, such as the Arduino Uno shown at the right in Fig. 1-6, you'll find a desktop computer more comfortable and convenient than a notebook. Get one with plenty of Universal Serial Bus (USB) ports, a microphone jack, a line input jack, and a headphone jack. Locate it on a table or desk separate from the one on which you do your "nuts-and-bolts" electronics construction and testing. That way, you'll avoid damaging your computer's keyboard or monitor when something flies away from your diagonal cutter or soldering gun in an unexpected direction!

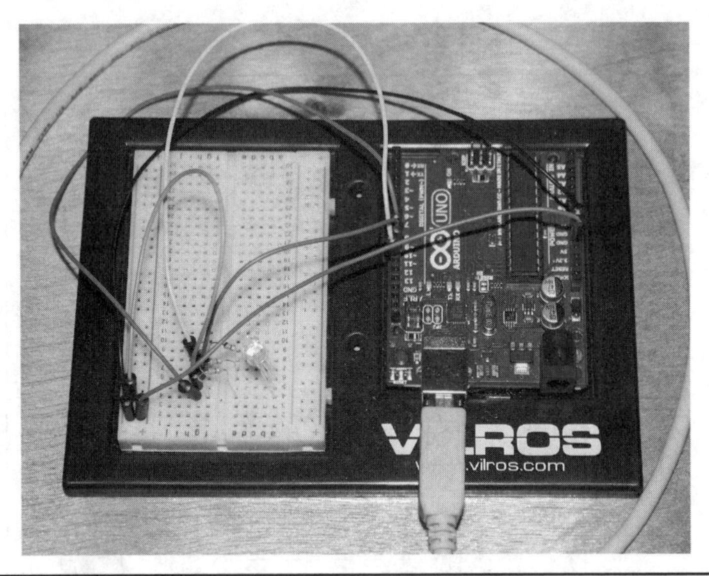

Figure 1-6 A miniaturized breadboard (at left) provided as part of a kit for use with an Arduino Uno microcomputer (at right).

The computer doesn't have to be a high-end machine; you're not likely to use resource-intensive software with it.

> **An Electronics Nerd's Apology**
>
> I have two desktop computers in my shop, a low-end one for the ham radio and a high-end one for video processing and production. I tend to go "over the top" when it comes to hobby electronics hardware.

Shortwave Radio

Shortwave listening (SWL) goes naturally along with hobby electronics. A modest shortwave receiver can "spice up" your shop, and might even lead you into more serious communications venues such as *Amateur Radio,* also called *ham radio*. You can buy a shortwave receiver ready-made, buy a kit that you can assemble into a receiver, or build your own radio from scratch.

The best commercially manufactured radios can run you upwards of a thousand dollars; the cheapest kits or "mini radios" will cost you only $30 or $40. If you want to get into SWL with a reasonably good radio right away, I recommend that you go to your local Radio Shack store and shop around. If they don't have physical units right on their shelves, go to the Radio Shack website, and you'll find a wide range of models, one of which will surely suit your desires! Their home page is at

www.radioshack.com

> **Tip**
>
> If you get interested in Amateur Radio, attend a meeting of your local ham radio club, or stop by one of the hundreds of ham radio conventions that take place nationwide. They'll provide all sorts of advice and insights.

Power Considerations

In a perfect universe, the alternating current (AC) electricity on your utility line would comprise a pure wave with no flaws or distortion. But in the real world, it's far from "clean." If you look at a typical utility AC waveform on a high-quality laboratory oscilloscope, you'll occasionally see voltage spikes called *transients* that greatly exceed the main wave's positive or negative peak voltage (Fig. 1-7). Transients can result from sudden changes in the *load* (the amount of power demanded by all the appliances combined) in an electrical circuit, and from various external disruptions to the utility hardware.

Coping with Transients

Lightning can, and often does, cause destructive transients on power lines. A single thundershower can produce frequent, intense transients over a large geographic area. Unless you take measures to suppress them, these spikes can destroy components in an electronic power supply. Transients can also interfere with the operation of sensitive equipment, such as computers or microcomputer-controlled appliances.

You can find commercially made *transient suppressors* in hardware stores and large department stores. These "surge protectors" use specialized semiconductor-based components to prevent sudden voltage spikes from reaching levels at which

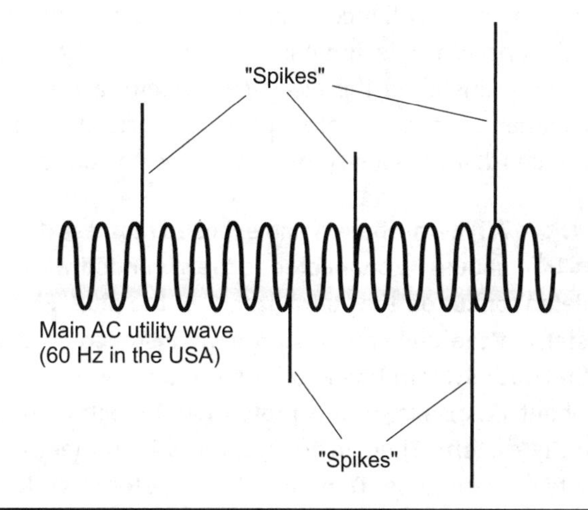

FIGURE 1-7 Transients appear as voltage "spikes" on an AC wave.

they can cause problems. The devices are rated in energy units known as *joules*, indicating the severity of the transients they can protect against.

In order to function, a transient suppressor requires a good electrical ground connection that, at some point, goes to a ground rod driven into the earth. Every residence or other building should have such a rod, preferably at the point where the utility line enters the structure. Transients can get directed away from sensitive equipment only when a current path exists for discharge to a good electrical ground. To guarantee such a connection, the building's wiring must have three-wire outlets, and the "third holes" in those outlets must lead to a well-designed electrical ground.

Tip

Use transient suppressors with all sensitive electronic equipment including computers, hi-fi stereo systems, and television sets (especially those expensive, big flat-screen ones). In the event of a thunderstorm, the best way to protect such equipment is to physically unplug every appliance from its wall outlet until the storm has passed.

Three-Wire System

In old buildings, you will often find two-wire AC systems. They have only two slots in the utility outlets. Some of these systems employ reasonable grounding by means of *polarization*, where one slot is longer than the other, and the longer slot goes to electrical ground. But that method never works as well as a three-wire AC system, in which the ground connection remains independent of both outlet slots.

Unfortunately, the presence of a three-wire or polarized outlet system doesn't always mean that an appliance connected to an outlet will actually be well-grounded. If the appliance design has faults, or if the "third slot" wasn't grounded by the people who installed the electrical system, a power supply can deliver a dangerous voltage to external metal appliance surfaces. This situation can pose an electrocution hazard, and can also hinder the performance of sensitive equipment.

Warning! **All metal chassis and exposed metal surfaces of AC power supplies should be connected to the grounded wire of a three-wire electrical cord. *Never* defeat or cut off the "third prong" of the plug. Find out whether or not the electrical system in the building was properly installed so that you don't live under the illusion that your system has a good ground when it really does not. If you have any doubts about this issue, hire a professional electrician to perform a complete inspection of the system. Then, if the system fails to meet code, get the work done as necessary to make it good. Don't wait for disaster to strike!**

All Tangled Up

Most of us have computer workstations, usually with multiple peripherals and ancillary equipment, such as a printer, a scanner, a modem, a router, a cordless phone, a desk lamp or two, a charging bay for devices such as tablet computers and cell phones, and so on. In the United States, all of these things get their power, either directly or indirectly, from a 117-V utility system. As a result, anyone with a substantial computer workstation will end up with a "tanglewire garden" behind and under the work desk. The same thing will happen if you gather enough gear in your electronics workshop.

"Tanglewire gardens" can look dangerous, as if they would present a high fire risk, but they needn't pose a hazard. If you know how to connect and arrange the wires properly, it doesn't matter from a safety standpoint how much you snarl them up, although you might want to affix labels on the cords near their end connectors so you don't get them confused when the inevitable malfunction occurs and you have to pull out and replace one of the components of your system.

Figure 1-8 shows the "tanglewire garden" underneath my ham radio station, which forms an extension of my electronics workshop. In addition to the radio itself, this system includes a computer, two displays, a digital communications interface between the radio and the computer, a microcomputer-controlled radio-frequency wattmeter, an audio amplifier for the computer and radio, a wireless headset, a desk lamp, and an external hard drive that needs its own "power brick." That's 10 devices or cords in total, all deriving their power from a single outlet in the wall protected by a 15-amp breaker at the main utility box.

Figure 1-8 "Tanglewire garden" beneath my electronics workbench. A heavy-duty UPS (out of the picture to the right) serves two power strips mounted on a metal baking sheet that rests on detached plastic shelves.

Keeping the Power Up

In order to ensure smooth operation of the system in case of a power failure, all of the devices go to the wall outlet through a commercially manufactured *uninterruptible power supply* (UPS). The UPS has a battery that charges from utility electricity under normal conditions, but provides a few minutes of emergency AC (with the help of some sophisticated electronic circuits) if the utility power fails, giving me time to deploy a small backup generator without having to shut any of the devices down. The UPS has four outlets in the back, two of which go to power strips with six outlets each, and the other two of which remain empty. There are 12 available outlets in the power strips, 10 of which are in active use. The UPS also has a transient suppressor built-in. Figure 1-9 is a block diagram of the arrangement.

Other Safety Measures

I've taken three extra precautions, aside from making sure that I don't overload the wall outlet, to ensure that my "tanglewire garden" remains safe.

1. First, if you look carefully at Fig. 1-8, you'll notice that I've mounted the power strips on an aluminum baking sheet. I glued the strips on the sheet with epoxy resin. This precaution keeps the power strips from setting anything (other than themselves) on fire if they short out.

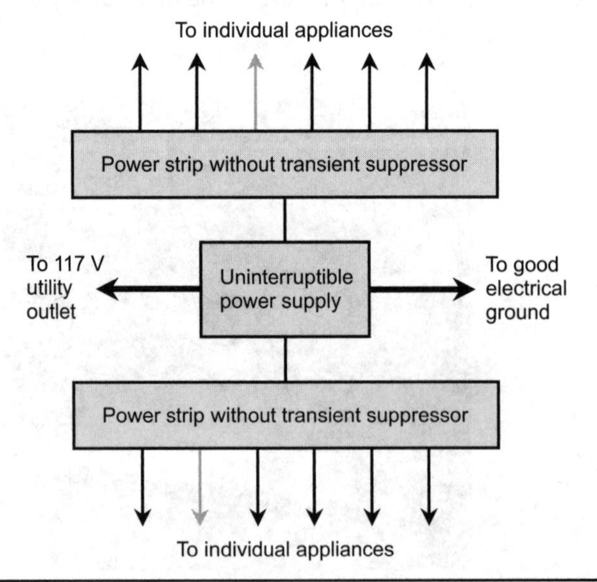

FIGURE 1-9 Block diagram of the "tanglewire garden" beneath the author's workbench. The power strips include breakers but not transient suppressors; the UPS contains a transient suppressor that serves the whole system. Gray arrows represent unused outlets.

2. Second, I don't let any cord splices or other sensitive electrical points lie directly on the floor. The baking sheets, as well as all points in the cords where splices exist, are set up on thick plastic shelves. Although I've never been flooded out, my basement floor will get wet if a sudden cloudburst occurs. (Of course, in that event I won't work in the shop until the floor dries out.)
3. Third, I've connected a ground wire from the chassis of the UPS to a known electrical ground. I tested the wall outlet underneath the workbench to ensure that the "third prong" actually goes to the electrical ground for the house.

Experiment: Dirty Electricity

You'll find this experiment easy to do with most computers. You'll need a 12-foot, two-wire audio cord with a 1/8-in two-conductor phone plug on one end and spade terminals on the other end (Radio Shack part number 42-2454). If you can't find that component, you can connect a 12-foot length of two-wire speaker cable to a 1/8-inch two-conductor phone plug.

Electromagnetic Fields

The current flowing through the utility grid produces obvious effects on appliances: glowing lamps, blowing fans, and chattering television sets. This AC also produces *electromagnetic* (EM) *fields* that aren't apparent to the casual observer. The presence of this EM energy causes tiny currents to flow or circulate in any object that conducts electricity, such as a wire, a metal rain gutter, the metal handle of your lawn mower, and your body. You can detect this energy and produce a multidimensional graphic display of its characteristics.

All EM fields display three independent properties: *amplitude*, *wavelength*, and *frequency*. The amplitude is the intensity of the field. The frequency is the number of full cycles per second. The wavelength is the distance in physical space between identical points on adjacent wavefronts. At 60 Hz, the AC utility frequency in the United States, EM waves are 5000 kilometers (approximately 3100 miles) long in *free space* (air or a vacuum).

The Software

You can use your computer to "look" at the EM fields permeating the atmosphere all around you. A simple computer freeware program called *DigiPan*, available on the Internet, can provide a real-time, moving graphical display of EM field components at frequencies ranging from 0 Hz (that is, DC) up to 5500 Hz. Here's the website:

www.digipan.net

DigiPan shows the frequency along the *x* axis (horizontally) and renders passing time as downward movement along the *y* axis (vertically). Figure 1-10 illustrates this display scheme in grayscale, but the real DigiPan display has colors. The relative intensity at each frequency appears as a specific hue, something like those weather radar images you've seen on television or the Internet. If there's no energy at a particular frequency, there's no line at all, and the display is black. If there's a little bit of energy at a particular frequency, you'll see a thin, vertical blue line creeping straight downward along the *y* axis. If there's a moderate amount of energy, the line turns yellow. If there's a lot of energy, the line becomes orange or red. The entire display is called a *waterfall*.

DigiPan is intended for digital communications in a mode called *phase-shift keying* (PSK). This mode is popular among amateur radio operators. You can read more about this interesting form of communications by "googling" it. Luckily for us, DigiPan can function as a very-low-frequency *spectrum monitor*, showing the presence of AC-induced EM fields not only at 60 Hz (which you should expect) but also at many other frequencies (which you might not expect until you see the evidence).

> **Tip**
>
> DigiPan doesn't take much computer processing power. Nearly all laptops or desktops can run it handily. If you have a good Internet connection, DigiPan will download and install in a minute or two.

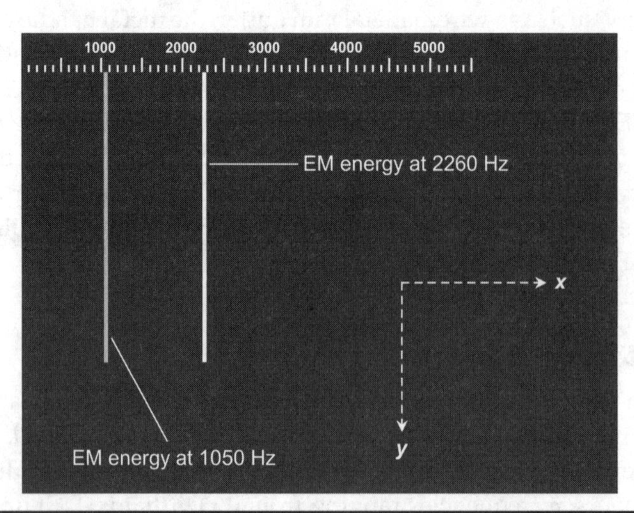

FIGURE 1-10. DigiPan display geometry. The horizontal or *x* axis portrays frequency. The vertical or *y* axis portrays time. The "signals" show up as steadily lengthening vertical lines. This drawing shows two hypothetical examples.

The Hardware

To observe the EM energy on your computer, you'll need an antenna. Cut off the spade lugs from the audio cord with a scissors or diagonal cutter. Separate the wires by pulling them apart along the entire length of the cord, so that you get a 1/8-inch two-conductor phone plug with two 12-foot wires attached.

Insert the phone plug into the *microphone* input of your computer. Arrange the two 12-foot wires so that they run in different directions from the phone plug. You can let the wires lie anywhere, as long as you don't trip over them! This arrangement will make the audio cord behave as a *dipole antenna* to pick up EM energy.

Open the audio control program on your computer. If you see a microphone input volume or sensitivity control, set it to maximum. If your audio program has a noise-reduction or noise-canceling feature, turn it off. Then launch DigiPan and carry out the following steps, in order.

- Click on "Options" in the menu bar and uncheck everything except "Rx."
- Click on "mode" in the menu bar and select "BPSK31."
- Click on "View" in the menu bar and uncheck everything.
- Click on "Configure" in the menu bar, select "Waterfall drive," select "Microphone," set the sliding balance control at the center, and maximize the volume.

Once you've completed these steps, the upper part of your computer display should show a chaotic jumble of text characters on a white background. The lower part of the screen should be black with a graduated scale at the top, showing numerals 1000, 2000, 3000, 4000, and 5000. Using your mouse, place the pointer on the upper border of the black region and drag that border upward until the white region with the distracting text vanishes.

If your little programmed gem works correctly, you should have a real-time, panoramic display of baseband EM energy from zero to several thousand hertz. Unless you're in a remote location far away from the utility grid, you should see a set of gradually lengthening vertical lines of various colors. These lines represent EM energy components at specific frequencies. You can read the frequencies from the graduated scale at the top of the screen. Do you notice a pattern?

Fundamental and Harmonics

A pure AC sine wave appears as a single *pip* or vertical line on the display of a spectrum monitor (Fig. 1-11A). This means that all of the energy in the wave is concentrated at one frequency, known as the *fundamental frequency*. But many, if not most, AC utility waves contain *harmonic* energy along with the energy at the fundamental frequency. Engineers sometimes refer to this phenomenon as *dirty electricity*.

A harmonic is a whole-number multiple of the fundamental frequency. For example, if 60 Hz is the fundamental frequency, then harmonics can exist at 120 Hz, 180 Hz, 240 Hz, and so on. The 120 Hz wave is the *second harmonic*, the 180 Hz wave is the *third harmonic*, the 240 Hz wave is the *fourth harmonic*, and so on. In general, if a wave has a frequency equal to n times the fundamental (where n is some whole number), then engineers call that wave the *nth harmonic*. In Fig. 1-11B, a wave is shown along with its second, third, and fourth harmonics, as the entire "signal" would appear on a spectrum monitor.

When you look at the EM spectrum display from zero to several thousand hertz using the arrangement described here, you'll see that utility AC energy contains not only the 60-Hz fundamental, but *many* harmonics. When I saw how much energy exists at the harmonic frequencies in my house, I could hardly believe it.

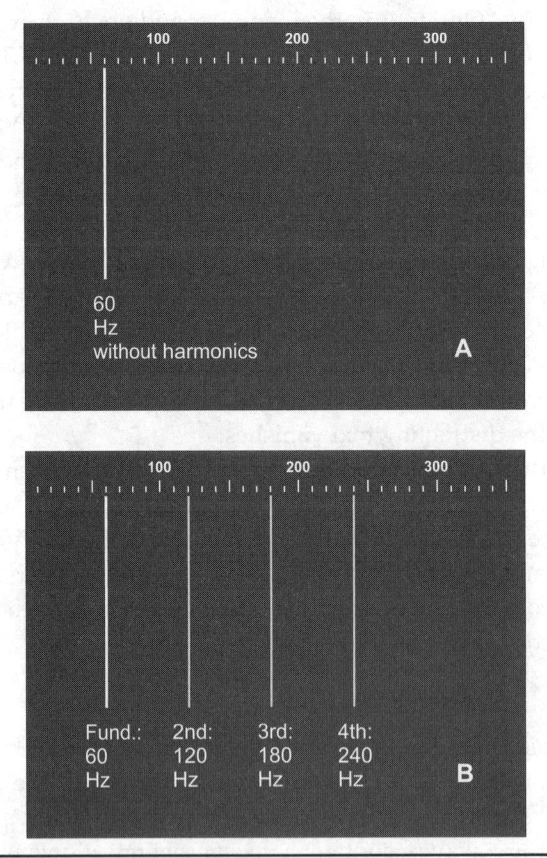

Figure 1-11 At A, a DigiPan display of pure, 60-Hz EM energy. At B, a display of 60-Hz energy with significant components at the second, third, and fourth harmonic (whole-number multiple) frequencies.

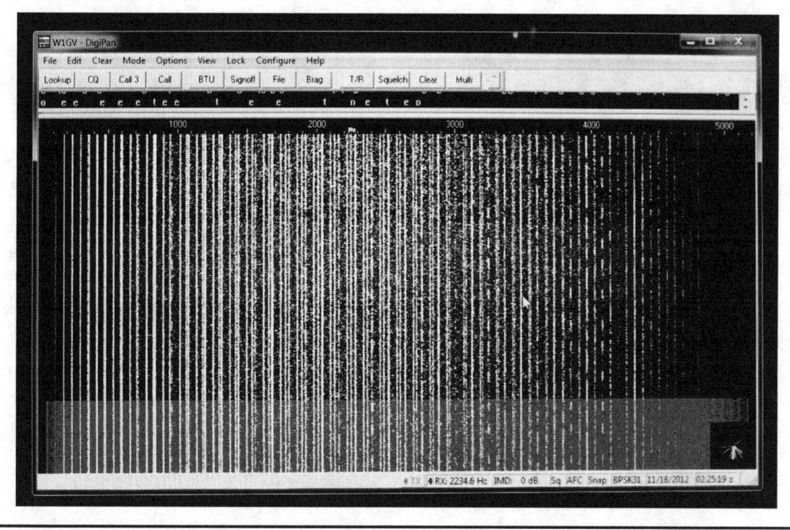

FIGURE 1-12 So-called "dirty electricity" in my work space, as viewed on DigiPan.

I had suspected some EM "grit," but not *that* much! Figure 1-12 shows an actual DigiPan display of dirty electricity in my shop. Each vertical trace represents an EM signal at a specific frequency. If the electricity were perfectly clean, we would see only one bright vertical trace at the extreme left end of the display.

Now Try This!

Place a vacuum cleaner near your EM pickup antenna. Power up the appliance while watching the DigiPan waterfall. When the motor first starts, do curves appear, veering to the right and then straightening out as vertical lines? Those contours indicate energy components that increase in frequency as the motor gets up to its operating speed, and maintain constant frequencies thereafter. When the motor loses power, do the vertical lines curve back toward the left before they vanish? Those curves indicate falling frequencies as the motor slows down. Try the same tests with a hair dryer, an electric can opener, or any other appliance with an electric motor. Which types of appliances make the most EM noise? Which make the least?

CHAPTER 2
Resistors

Resistors oppose the flow of electric current through circuits. They come in many shapes and sizes, manufactured for a wide variety of purposes.

Fixed Resistors

In your electronics lab, you'll often use *fixed resistors*, components that maintain a constant opposition to electric current, converting some of the current that passes through them to heat energy.

Carbon-Composition Resistors

The cheapest way to make a resistor involves mixing powdered carbon with nonconducting paste, pressing the "goo" into a cylindrical shape, inserting wires in the ends, and then letting the mass harden (Fig. 2-1). This process yields a *carbon-composition resistor* that introduces no inductance or capacitance into a circuit, but only resistance.

Carbon-composition resistors dissipate power in proportion to their physical size and mass. Most of the carbon-composition resistors sold today can handle 1/4 watt (W) or 1/2 W. You can also find 1/8-W units for use in miniaturized, low-power circuitry, and 1-W or 2-W units for circuits that require electrical ruggedness. Occasionally you'll see a carbon-composition resistor with a power rating such as 50 W or 60 W.

Wirewound Resistors

An alternative method of making a resistor involves wrapping a length of poorly conducting wire around a rod of insulating material (Fig. 2-2). The resistance depends on how well the wire conducts, on its diameter or *gauge*, and on its length. When manufacturers construct a component in this fashion, they end up with a *wirewound resistor*.

Some wirewound resistors can handle large amounts of power. That's their best asset. On the downside, wirewound resistors exhibit inductance as well as

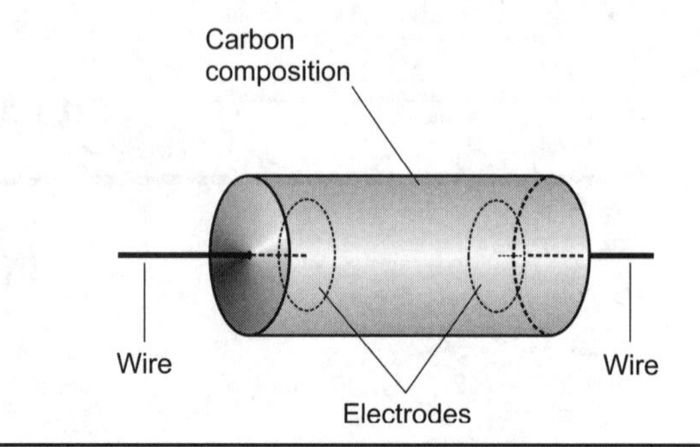

Figure 2-1 Construction of a carbon-composition resistor.

resistance, making them a poor choice for use in situations where high-frequency *alternating-current* (AC) or *radio-frequency* (RF) energy exists.

> **Tip**
>
> You'll find wirewound resistors whose actual resistance values, expressed and measured in units called *ohms*, fall within a narrow range (sometimes a fraction of 1 percent either way of the quoted number). Engineers say that such components have *close tolerance* or *tight tolerance*.

Film Type Resistors

Manufacturers can apply carbon paste, resistive wire, or some mixture of ceramic and metal to a cylindrical form as a film or thin layer to obtain a specific resistance. When they do this, they get a *carbon-film resistor* or *metal-film resistor*. The component looks like a carbon-composition resistor from the outside, but the internal construction

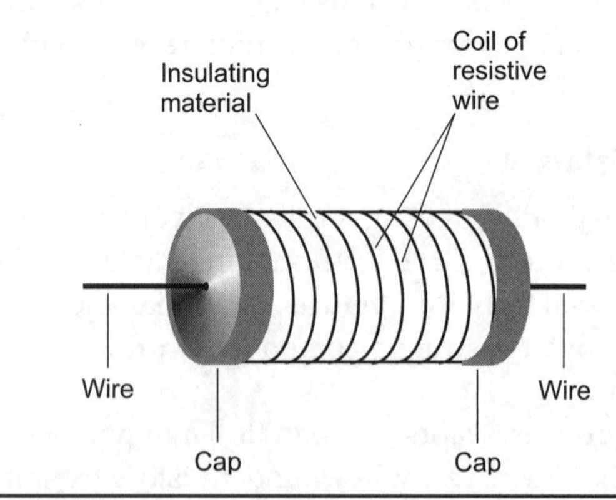

Figure 2-2 Construction of a wirewound resistor.

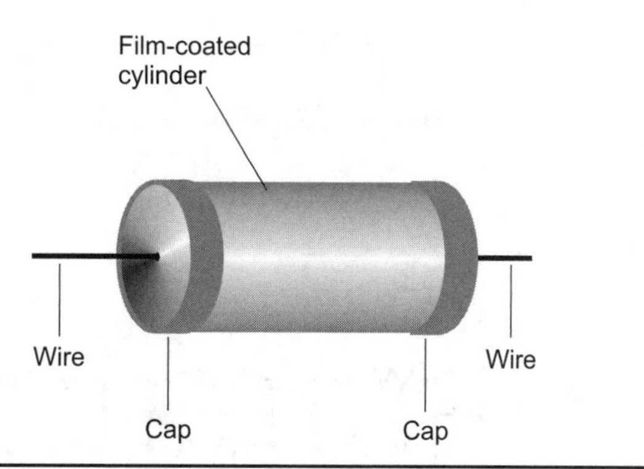

Figure 2-3 Construction of a film type resistor.

differs (Fig. 2-3). Film type resistors, like carbon-composition resistors, have little or no inductance, but in general they can't handle as much power as carbon-composition or wirewound types of comparable physical size.

Power Ratings

A resistor bears a specification that tells you how much power it can safely *dissipate* (turn it into heat). The rating applies to *continuous duty* operation, meaning that the component can dissipate a certain amount of power constantly and indefinitely.

Calculating the Allowable Current

You can calculate how much current a given resistor can handle by using the formula for power P (in watts) in terms of current I (in amperes) and resistance R (in ohms), as follows:

$$P = I^2R$$

With algebra, you can change this formula to express the maximum allowable current in terms of the power dissipation rating and the resistance:

$$I = (P/R)^{1/2}$$

In engineering mathematics, a "1/2 power" notation represents the positive square root.

Resistor Matrices

You can multiply resistor power ratings by connecting several of them in a *square matrix* of 2 × 2, 3 × 3, 4 × 4, or larger. A matrix of n by n resistors, all having *identical* ohmic values and *identical* power ratings, has n^2 times the power-handling capability

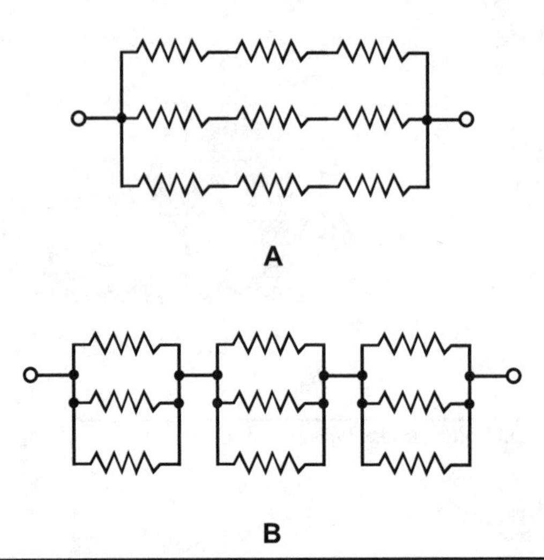

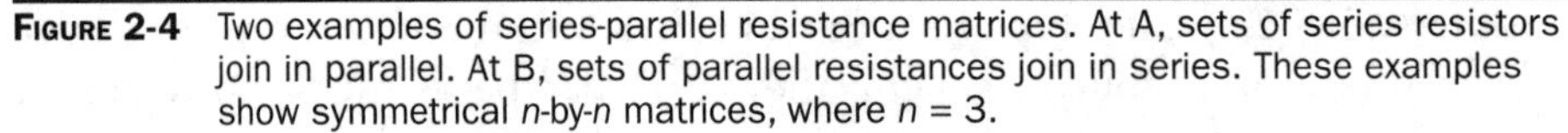

FIGURE 2-4 Two examples of series-parallel resistance matrices. At A, sets of series resistors join in parallel. At B, sets of parallel resistances join in series. These examples show symmetrical *n*-by-*n* matrices, where $n = 3$.

of any one resistor alone. For example, a 3-by-3 matrix of 1-W resistors can handle 3^2 or 9 W.

Figure 2-4 shows two ways of getting a 3-by-3 resistor matrix. At A, you assemble three sets of three identical resistors in series (like the links in a chain), and then connect those three sets in parallel (like the rungs in a ladder). At B, you assemble three sets of three identical resistors in parallel, and then connect those three sets in series.

Non-symmetrical, series-parallel networks, made up from identical resistors, can increase the power-handling capability over that of a single resistor. But in these cases, the total resistance differs from the value of any individual resistor. To obtain the overall power-handling capacity, you can always multiply the power-handling capacity of any individual resistor by the total number of resistors, whether the network is symmetrical or not—again, *if and only if,* all the resistors have identical ohmic values and identical power-dissipation ratings.

Tip

Resistor dissipation ratings, like the ohmic values, are specified with a margin for error. A good engineer never tries to "push the rating" and use, say, a 1-W resistor in a situation where it will draw 1.2 W. In fact, good engineers usually include their own safety margin, in addition to that offered by the vendor. For example, if you are allowing for a 10% safety margin, you wouldn't demand that a 9-W resistor handle more than 90% of 9 W, or 8.1 W.

Ohmic Values

In theory, a resistor can have any ohmic value from the lowest possible (such as a shaft of solid silver) to the highest (dry air). In practice, you'll rarely find resistors with values less than 0.1 ohm (Ω) or more than 999 megohm (M).

Resistors are manufactured with ohmic values in power-of-10 multiples of numbers from the set

{1.0, 1.2, 1.5, 1.8, 2.2, 2.7, 3.3, 3.9, 4.7, 5.6, 6.8, 8.2}

You'll routinely see resistances, such as 47 ohms, 180 ohms, 6.8 kilohms (k), or 18 M, but you'll hardly ever find resistors with values, such as 384 ohms, 4.54 k, or 7.297 M.

Additional basic resistances exist, intended especially for *tight-tolerance* (or *precision*) resistors: power-of-10 multiples of numbers from the set

{1.1, 1.3, 1.6, 2.0, 2.4, 3.0, 3.6, 4.3, 5.1, 6.2, 7.5, 9.1}

Tolerance

The first set of numbers above represents standard resistance values available in tolerances of plus or minus 10 percent (±10%), meaning that the resistance might be as much as 10% more or less than the indicated amount. In the case of a 470-ohm resistor, for example, the value can be larger or smaller than the rated value by as much as 47 ohms, and still adhere to the rated tolerance. That's a range of 423 to 517 ohms.

Engineers calculate resistor tolerance figures on the basis of the *rated* resistance, not the measured resistance. For example, you might test a "470-ohm" resistor and find it to have an actual resistance of 427 ohms. This discrepancy would leave the component within ±10% of the specified value. But if you test it and find it to have a resistance of 420 ohms, its actual value falls outside the rated range, so it constitutes a "reject."

The second set of numbers listed above, along with the first set, represents all standard resistance values available in tolerances of plus or minus 5 percent (±5%). A 470-ohm, 5% resistor will have an actual value of 470 ohms plus or minus 24 ohms, or a range of 446 to 494 ohms.

For applications requiring exceptional precision, resistors exist that boast tolerances tighter than ± 5%. You might need a resistor of such quality in a circuit or system where a small error can make a big difference.

Color Codes

Some resistors have *color bands* that indicate ohmic value and tolerance. You'll see three, four, or five bands around carbon-composition resistors and film resistors.

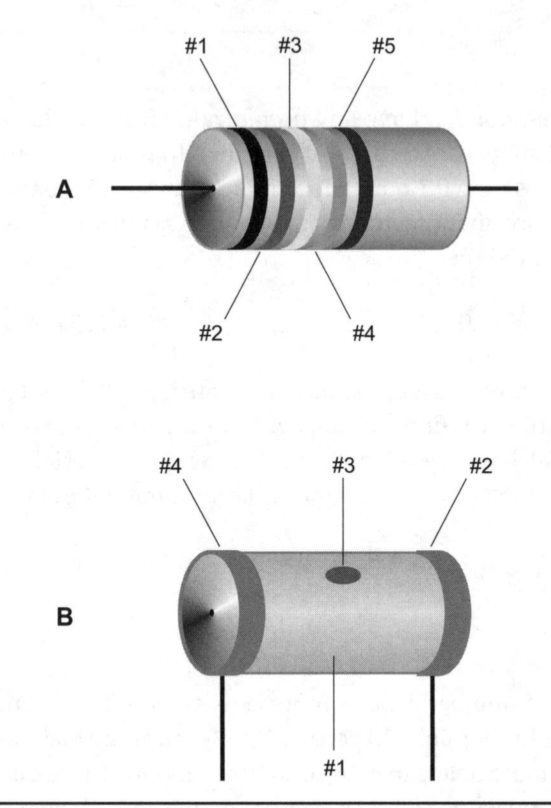

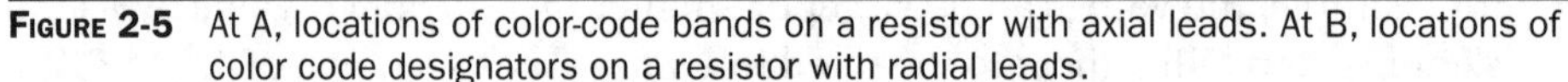

FIGURE 2-5 At A, locations of color-code bands on a resistor with axial leads. At B, locations of color code designators on a resistor with radial leads.

Other resistors are big enough for printed numbers that disclose the values and tolerances.

On resistors with *axial leads* (wires that come straight out of both ends), the first, second, third, fourth, and fifth bands are arranged as shown at A in Fig. 2-5. On resistors with *radial leads* (wires that come off the ends at right angles), the colored regions are arranged as shown in Fig. 2-5B. The first two regions represent single digits 0 through 9, and the third region represents a multiplier of 10 to some power. (For the moment, don't worry about the fourth and fifth regions.) Table 2-1 indicates the numerals corresponding to various colors.

Suppose that you find a resistor with three bands: yellow, violet, and red, in that order. You can read as follows, from left to right, referring to Table 2-1:

- Yellow = 4
- Violet = 7
- Red = ×100

Table 2-1 Color codes for the first three bands that appear on fixed resistors. See text for discussion of the fourth and fifth bands

Color of band	Numeral (first and second bands)	Multiplier (third band)
Black	0	1
Brown	1	10
Red	2	100
Orange	3	1000 (1 k)
Yellow	4	10^4 (10 k)
Green	5	10^5 (100 k)
Blue	6	10^6 (1 M)
Violet	7	10^7 (10 M)
Gray	8	10^8 (100 M)
White	9	10^9 (1000 M or 1 G)

You conclude that the rated resistance equals 4700 ohms, or 4.7 k. As another example, suppose you find a resistor with bands of blue, gray, and orange. You refer to Table 2-1 and determine that

- Blue = 6
- Gray = 8
- Orange = ×1000

This sequence tells you that the resistor is rated at 68,000 ohms, or 68 k.

If a resistor has a fourth colored band on its surface (#4 as shown in Fig. 2-5 A or B), then that mark tells you the tolerance. A silver band indicates ±10%. A gold band indicates ±5%. If no fourth band exists, then the tolerance is ±20%.

The fifth band, if any, indicates the maximum percentage by which you should expect the resistance to change after the first 1000 hours of use. A brown band indicates a maximum change of ±1% of the rated value. A red band indicates ±0.1%. An orange band indicates ±0.01%. A yellow band indicates ±0.001%. If the resistor lacks a fifth band, it tells you that the resistor might deviate by more than ±1% of the rated value after the first 1000 hours of use.

Tip

Always test a resistor with an ohmmeter before installing it in a circuit. It takes only a few seconds to check a resistor's value. In contrast, once you've finished building a circuit and discover that it won't work because of some miscreant resistor, the troubleshooting process can take hours.

Variable Resistors

Figure 2-6A illustrates the construction geometry of a variable resistor called a *potentiometer*. Figure 2-6B shows the schematic symbol. If a potentiometer uses a resistance wire rather than solid strip resistive material, it's called a *rheostat*.

How They Work

In a conventional potentiometer, a strip of resistive film goes 3/4 of a circle around (an arc of 270°), with terminals connected to the ends. To obtain variable resistance, a sliding contact, attached to a rotatable shaft and bearing, goes to the middle terminal. The resistance between the middle terminal and either end terminal can, therefore, vary from zero up to the resistance of the whole strip.

Linear-Taper Potentiometer

A *linear-taper potentiometer* uses a strip of resistive material with constant density all the way around. As a result, the resistance between the center terminal and either end terminal changes at a steady rate as the control shaft rotates. Engineers usually prefer linear-taper potentiometers in electronic test instruments. Linear-taper potentiometers also exist in some consumer electronic devices.

Audio-Taper Potentiometer

Humans perceive sound intensity according to the *logarithm* of the actual sound power, not in direct proportion to the actual power. An *audio-taper potentiometer* can compensate for this effect. The resistance varies according to a *nonlinear function* of

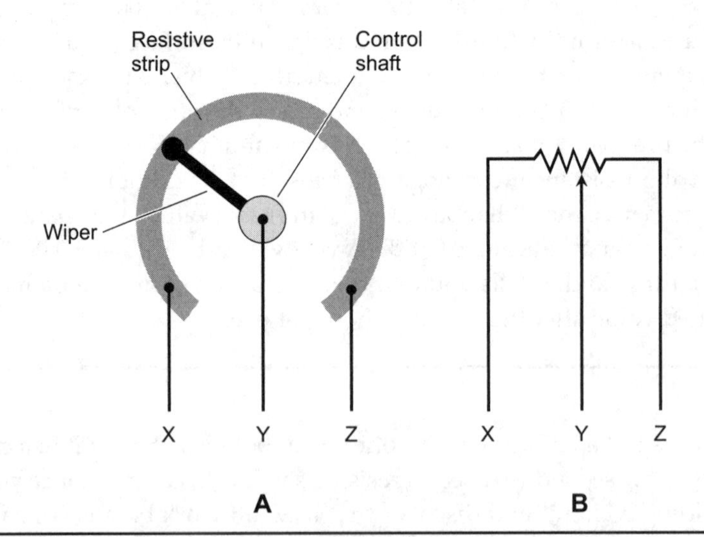

FIGURE 2-6 Functional drawing of a rotary potentiometer (A) and the schematic symbol for the same component (B).

the angular shaft position because the resistive strip doesn't have constant density all the way around.

> **Tip**
>
> Some engineers call audio-taper potentiometers *logarithmic-taper* or *log-taper potentiometers* because the function of resistance versus rotational position makes a logarithmic curve when you graph it.

Slide Potentiometer

A potentiometer can employ a straight strip of resistive material rather than a circular strip so that the control moves up and down, or from side to side, in a straight line. This type of variable resistor, called a *slide potentiometer*, finds applications in hi-fi audio *graphic equalizers*, as volume controls in some amplifiers, and in other applications where operators prefer a straight-line control movement to a rotating control movement. Slide potentiometers exist in both linear-taper and audio-taper configurations.

Rheostat

A variable resistor can employ a wirewound element, rather than a solid strip of resistive material. Engineers call this type of device a *rheostat*. It can have either a rotary control or a sliding control. You can't adjust a rheostat in a perfectly smooth manner as you can do with a potentiometer because the movable contact slides along the wire coil from a certain point on one turn to the adjacent point on the next turn. The smallest possible increment of resistance equals the amount of resistance in one wire turn.

Handy Math

Figure 2-7 illustrates a generic circuit with a variable direct-current (DC) generator, a voltmeter, some wire, an ammeter, and a potentiometer. Voltage (in volts) is represented as E, current (in amperes) as I, and resistance (in ohms) as R.

Ohm's Law Formulas

To calculate the voltage when you know the current and the resistance in a DC circuit, such as the one shown in Fig. 2-7, use the formula

$$E = IR$$

To calculate the current when you know the voltage and the resistance, use

$$I = E/R$$

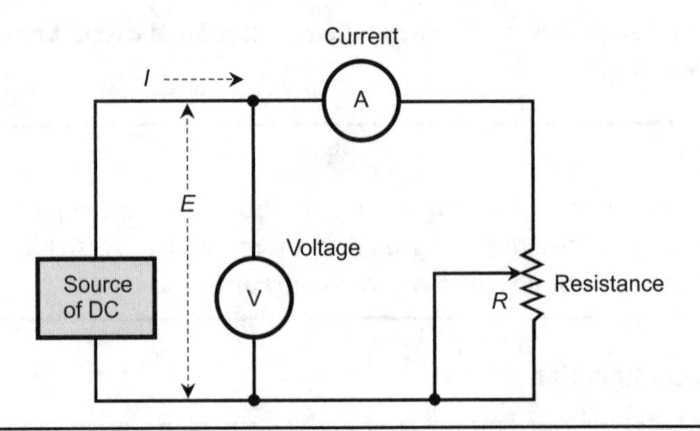

FIGURE 2-7 A circuit for doing DC circuit calculations. The voltage (in volts) is E, the current (in amperes) is I, and the resistance (in ohms) is R.

To calculate the resistance when you know the voltage and the current, use

$$R = E/I$$

If you want Ohm's Law to produce the correct results, you must input the proper units to the formulas. Under most circumstances, you'll use the *standard units* of volts (V), amperes (A), and ohms (Ω).

Power Formulas

You can calculate the power P (in watts) in a DC circuit, such as the one shown in Fig. 2-7, using the formula

$$P = EI$$

If you know I and R but don't know E, then you can get the power P as

$$P = I^2R$$

If you know E and R but don't know I, then you can calculate the power P as

$$P = E^2/R$$

Combining Resistors

When you place resistances in series, their ohmic values add up to get the total (or *net*) resistance. If the resistance values are R_1, R_2, R_3, and so on up to R_n, then the net resistance R of the whole set is

$$R = R_1 + R_2 + R_3 + \ldots + R_n$$

You can calculate the resistance of a set of resistors connected in parallel according to the formula

$$R = 1/(1/R_1 + 1/R_2 + 1/R_3 + \ldots + 1/R_n)$$

> **Tip**
> Once in a while, you'll encounter a situation where multiple resistances appear in parallel and their values are all equal. In that case, the total resistance is the resistance of any one component divided by the number of components.

Division of Power

When you connect sets of resistors to a source of voltage, each resistor draws some current. If you know the source voltage, you can figure out how much current the entire set consumes by calculating the net resistance of the combination, and then considering the combination as a single resistor.

If the resistors in the network all have the same ohmic value, then the power from the source divides up equally among them, whether you connect the resistors in series or parallel. For example, if you have eight identical resistors in series with a battery, the network consumes a certain amount of power, each resistor bearing 1/8 of the load. If you rearrange the circuit to connect the resistors in parallel with the same battery, the network *as a whole* dissipates a lot more power, but each *individual* resistor handles 1/8 of the total power load, just as they do when they're in series. If the resistances in a network do not all have identical ohmic values, then some resistors dissipate more power than others.

Experiment 1: Resistance of a Liquid

In this experiment, you'll measure the resistance of plain tap water, and then add salt to increase the conductivity.

Resistance of Plain Tap Water

Connect both of your meter probe wires to copper electrodes, which you can fabricate from pipe clamps available at hardware stores by pounding the clamps flat with a hammer. Plug the negative (black) meter probe jack into the common-ground meter input, but leave the positive (red) probe jack unplugged. Place the two copper electrodes into a glass cup full of water. Make sure that the strips rest on opposite sides of the cup, so they're as far away from each other as possible. Bend the strips over the edges of the cup to hold them in place, as shown at A in Fig. 2-8. Fill the cup with water until the liquid comes almost up to the brim. Use tap water, not distilled or bottled water.

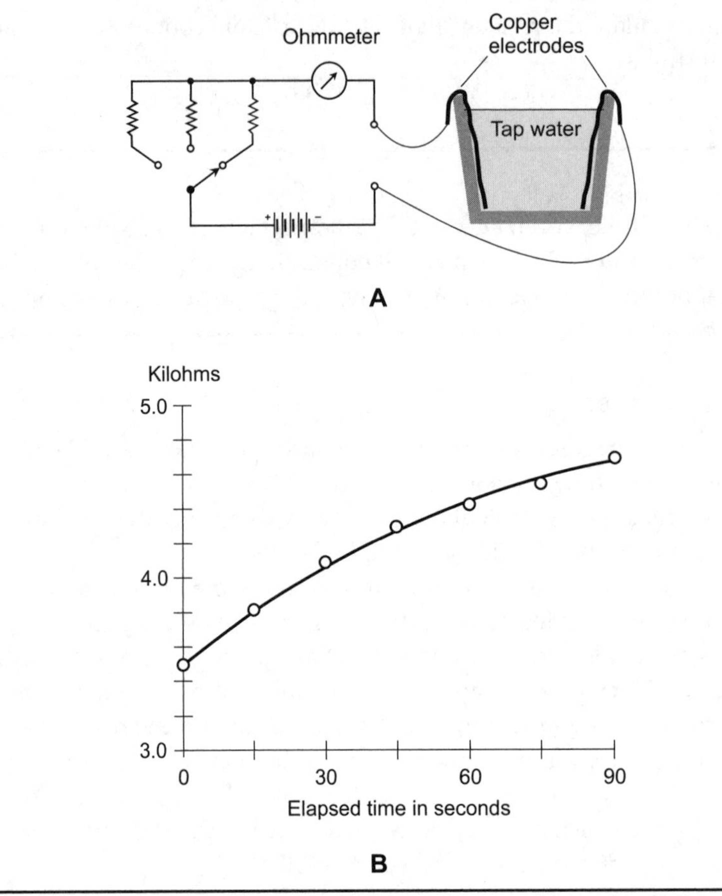

FIGURE 2-8 At A, an arrangement for measuring the resistance of tap water. At B, resistance values as I measured them over a time span of 90 seconds.

Switch your ohmmeter to measure resistance in a range from 0 to several kilohms. Find a clock or watch with a second hand, or a digital clock or watch that displays seconds as they pass. When the second hand reaches the "top of the minute" or the digital seconds display indicates "00," plug the positive meter probe jack into its receptacle. Note the resistance at that moment. Then, keeping one eye on the clock and the other eye on the ohmmeter display, note and record the resistances at 15-second intervals until 90 seconds have elapsed. When I did this experiment, I got the following results.

- At the beginning: 3.49 kilohms
- After 15 seconds: 3.82 kilohms
- After 30 seconds: 4.10 kilohms
- After 45 seconds: 4.30 kilohms

- After 60 seconds: 4.42 kilohms
- After 75 seconds: 4.54 kilohms
- After 90 seconds: 4.68 kilohms

Figure 2-8B graphically shows these results. The small open circles represent actual data points. The black curve is an optimized graph obtained by *curve fitting*. Your results will differ from mine depending on the type of ohmmeter you use, the dimensions of your electrodes, and the mineral content of your tap water. In any case, you should observe a gradual increase in the resistance of the water as time passes.

Resistance of Salt Water

Remove the positive meter probe jack from its meter receptacle. Switch the ohmmeter to the *next lower* resistance range. Carefully measure out 0.5 teaspoon of table salt. Use a measuring spoon (available at good grocery stores) for this purpose, and level off the salt to ensure that the amount comes as close to 0.5 teaspoon as possible. Pour the salt into the water. Stir the solution until the salt has completely dissolved. Then allow the solution to settle down for a minute.

Once again, watch the clock. At the "top of the minute," plug the positive meter probe jack into its receptacle. Your system should now be interconnected as shown at A in Fig. 2-9. Record the resistances at 15-second intervals. Here are the results I got.

- At the beginning: 250 ohms
- After 15 seconds: 262 ohms
- After 30 seconds: 269 ohms
- After 45 seconds: 273 ohms
- After 60 seconds: 276 ohms
- After 75 seconds: 282 ohms
- After 90 seconds: 286 ohms

Remove the positive meter probe jack from its outlet. Add another 0.5 teaspoon of salt to the solution, stir it in until it's totally dissolved, and then let the solution settle down. Repeat your timed measurements. I got the following results.

- At the beginning: 165 ohms
- After 15 seconds: 193 ohms
- After 30 seconds: 201 ohms
- After 45 seconds: 207 ohms
- After 60 seconds: 213 ohms
- After 75 seconds: 219 ohms
- After 90 seconds: 224 ohms

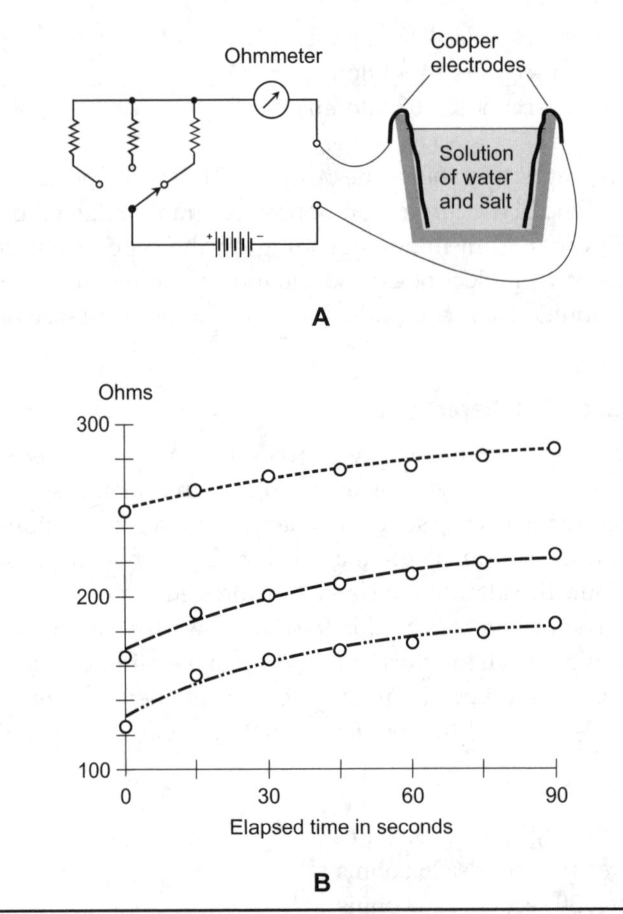

FIGURE 2-9 At A, an arrangement for measuring the resistance of water with salt fully dissolved. At B, resistances I observed over a span of 90 seconds. Upper (short-dashed) curve: 0.5 teaspoon of salt. Middle (long-dashed) curve: 1.0 teaspoon of salt. Lower (dashed-and-dotted) curve: 1.5 teaspoons of salt.

Once again, disconnect the positive probe wire from the meter. Add a third 0.5 teaspoon of salt to the solution, stir until it's dissolved, and let the solution settle. Do another series of timed measurements. Here are my results.

- At the beginning: 125 ohms
- After 15 seconds: 156 ohms
- After 30 seconds: 163 ohms
- After 45 seconds: 168 ohms
- After 60 seconds: 173 ohms
- After 75 seconds: 179 ohms
- After 90 seconds: 185 ohms

You should observe, as I did, a general decrease in the solution's resistance as the salt concentration goes up, and an increase in the resistance over time with the ohmmeter connected. Figure 2-9B is a multiple-curve graph of the data tabulated above.

Why Does the Resistance Rise?

In these experiments, *electrolysis* occurs as current passes through the solution. In electrolysis, water (H_2O) molecules break apart into hydrogen (H) and oxygen (O), both of which are gases at room temperature. The gases accumulate as bubbles on the electrodes, some of which rise to the surface of the solution. However, both electrodes remain "coated" with some bubbles, which effectively reduce the surface area of metal in contact with the liquid. That's why the apparent resistance of the solution goes up over time. If you stir the solution, you'll knock the bubbles off of the electrodes for a few moments, and the measured resistance will drop back down. If you let the solution come to rest again, the apparent resistance will rise once more as new gas bubbles accumulate on the electrodes.

Experiment 2: Resistors in Series

In this experiment, you'll see how the ohmic values of resistors combine when you connect them in series. You'll need a 1.5-V flashlight cell, a holder for the cell, your multimeter, and 1/2-W resistors with manufacturer-quoted values of 330 ohms, 1000 ohms, and 1500 ohms. You can get all these parts at Radio Shack.

Test A

When I measured the ohmic values for the resistors rated at 330, 1000, and 1500 ohms, I got 326 ohms, 981 ohms, and 1467 ohms, respectively. (Your resistors will have slightly different values, most likely.) When I connected the 326-ohm resistor in series with the 981-ohm resistor, as shown in Fig. 2-10, I expected to get a total resistance of

$$R = 326 + 981 = 1307 \text{ ohms}$$

I twisted the resistors' leads together and then measured the total series resistance by holding the meter probe tips against the outer leads. I wore gloves to keep my body resistance from affecting the reading. With my digital ohmmeter set to the range 0 to 2000 ohms, I got a display reading of 1305 ohms: a slight error, but entirely attributable to the imperfect nature of the physical universe!

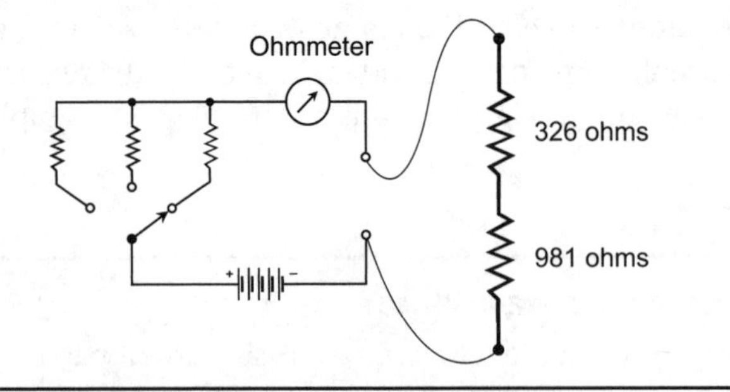

FIGURE 2-10 Series combination of 326 ohms and 981 ohms.

Test B

For the second part of the experiment, I connected my 326-ohm resistor in series with my 1467-ohm resistor, as shown in Fig. 2-11, again twisting the leads together and holding the meter probe tips against the outer leads. I expected to get a total resistance of

$$R = 326 + 1467 = 1793 \text{ ohms}$$

My meter, again set for the range 0 to 2000 ohms, displayed 1791 ohms. I was satisfied, once again, that the laws of basic electricity were being obeyed. Your smallest and largest resistors should behave in the same orderly fashion when you connect them in series and measure the total resistance.

Test C

Phase 3 of the experiment involved the two largest resistors: the 981-ohm component and the 1467-ohm component. When I twisted the leads together and measured the total resistance, as shown in Fig. 2-12, I expected to obtain

$$R = 981 + 1467 = 2448 \text{ ohms}$$

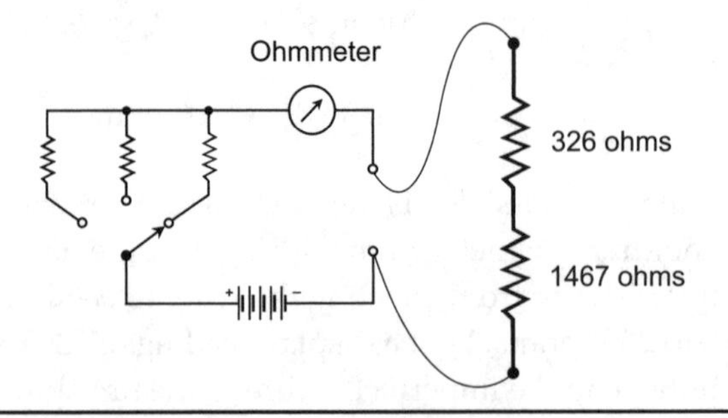

FIGURE 2-11 Series combination of 326 ohms and 1467 ohms.

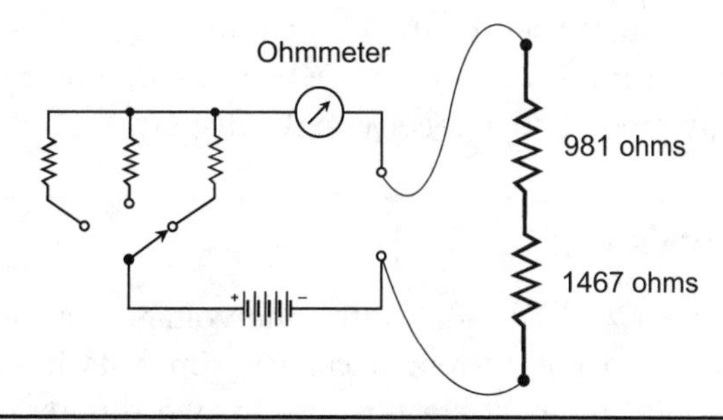

FIGURE 2-12 Series combination of 981 ohms and 1467 ohms.

My digital ohmmeter won't display this value when set for the range 0 to 2000 ohms, so I set it to the next higher range, which was 0 to 20 kilohms. The display showed 2.44, meaning 2.44 kilohms (2.44K ohms). My predicted value, rounded off to the nearest hundredth of a kilohm, was 2.45K ohms.

Test D

For the final part of this experiment, I connected all three of the resistors in series, in order of increasing resistance, by twisting the leads together. I placed the 326-ohm resistor on one end, the 981-ohm resistor in the middle, and the 1467-ohm resistor on the other end. Then I connected the ohmmeter as shown in Fig. 2-13. According to my prediction, the total resistance should have been

$$R = 326 + 981 + 1467 = 2774 \text{ ohms}$$

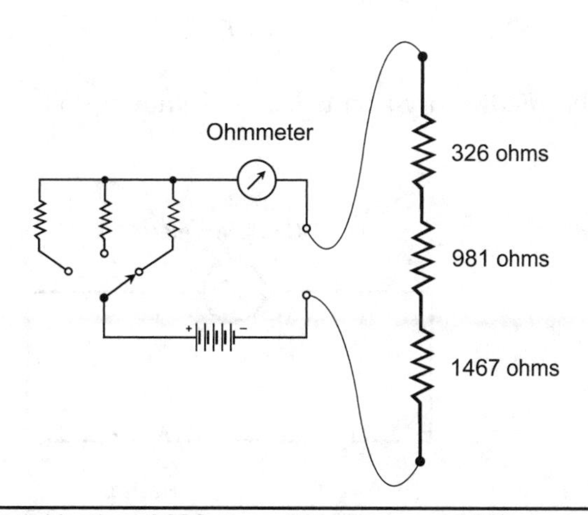

FIGURE 2-13 Series combination of 326 ohms, 981 ohms, and 1467 ohms.

With the ohmmeter set for the range 0 to 20K, I got a reading of 2.77 on the digital display, meaning 2.77K ohms. When rounded off to the nearest hundredth of a kilohm, my prediction agreed with the display.

Experiment 3: Ohm's Law

In its basic form, *Ohm's law* states that the voltage across a component varies in direct proportion to the current it carries, times its internal resistance. In this experiment, you'll demonstrate this law in two different ways. You'll need the same components as you used in the previous experiment.

Check the Components

As a lifelong electronics nerd, I recommend that you test every component *before* you put it to work in a circuit or system. If you check a component in isolation and find it bad, toss it aside and replace it. If you let a defective component become part of a circuit, it can create problems that you might find impossible to track down without taking the whole circuit apart.

When I tested my AA cell with my digital voltmeter, I got a reading of 1.585 V. When I tested the resistors with my digital ohmmeter in the previous experiment, I observed 326 ohms, 981 ohms, and 1467 ohms. All of these values came within the manufacturer's rated specifications. If they hadn't, I would have discarded and replaced the defective one(s).

Test A: Current Measurement

Once you have determined the actual voltage and resistances, you're ready to predict how much current the cell will deliver through each of the three resistors individually. Figure 2-14 shows the arrangement for making the measurements. Use the Ohm's law formula

$$I = E/R$$

where I is the predicted current, E is the known voltage, and R is the known resistance.

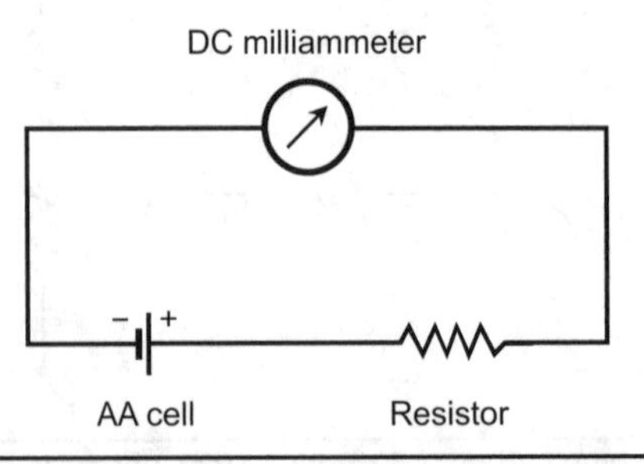

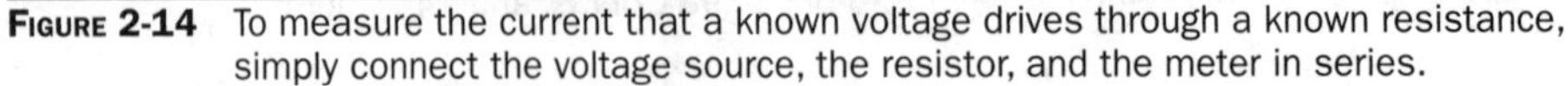

FIGURE 2-14 To measure the current that a known voltage drives through a known resistance, simply connect the voltage source, the resistor, and the meter in series.

With the 326-ohm resistor connected in series with the cell and the meter, I predicted that I would see a current value of

$$\begin{aligned} I &= 1.585/326 \\ &= 0.00486 \text{ A} \\ &= 4.86 \text{ mA} \end{aligned}$$

When I made the measurement, I got 4.69 milliamperes, which was 0.17 milliampere below the predicted current.

With the 981-ohm resistor connected in series with the cell and the meter, I predicted that I would observe

$$\begin{aligned} I &= 1.585/981 \\ &= 0.00162 \text{ A} \\ &= 1.62 \text{ mA} \end{aligned}$$

I measured an actual current of 1.60 mA, which was 0.02 mA below the predicted value.

With the 1467-ohm resistor connected in series with the cell and the meter, I predicted that I would see

$$\begin{aligned} I &= 1.585/1467 \\ &= 0.00108 \text{ A} \\ &= 1.08 \text{ milliamperes} \end{aligned}$$

I measured 1.08 mA, exactly the predicted value.

Experimental Error

In any laboratory environment, you will have to deal with some *experimental error*. No instrument lives up to theoretical ideals, and the physical world is, in a certain sense, "fuzzy" by nature. To determine the error as a percentage (%) of the predicted value, do these calculations in order:

- Take the measured value
- Subtract the predicted value
- Divide by the predicted value
- Multiply by 100
- Round the answer off as needed

Make sure that you use the same units for the measured and predicted values. In the first case, my experimental error worked out as

$$\begin{aligned} &[(4.69 - 4.86)/4.86] \times 100 \\ &= (-0.17/4.86) \times 100 \\ &= -0.0350 \times 100 \\ &= -3.50\% \end{aligned}$$

which rounds off to −4%. In the second case, my experimental error was

$$\begin{aligned}&[(1.60 - 1.62)/1.62] \times 100\\ &= (-0.02/1.62) \times 100\\ &= -0.0123 \times 100\\ &= -1.23\%\end{aligned}$$

which rounds off to −1%. In the third case, my experimental error was

$$\begin{aligned}&[(1.08 - 1.08)/1.08] \times 100\\ &= (0.00/1.08) \times 100\\ &= 0.00 \times 100\\ &= 0.00\%\end{aligned}$$

which rounds off to 0%. You can expect to get errors of a magnitude similar to those I experienced, either positive or negative.

Tip

If you use an analog meter, you must contend with *interpolation error* as well as experimental error. That's because you'll have to *interpolate* (make a good guess at) the position of the needle on the meter scale.

Test B: Voltage Measurement

The second Ohm's law experiment involves measuring the voltages across each of the three resistors when you connect them in series with the cell across the combination, as shown in Fig. 2-15. Here, you should use the Ohm's law formula

$$E = IR$$

where E is the predicted voltage, I is the measured current, and R is the measured resistance.

When I connected the three resistors in series with the DC milliammeter and the cell, I observed a current of 553 microamperes through the combination of resistors, equivalent to 0.000553 ampere. Across the 326-ohm resistor (Fig. 2-15A), I predicted that I would see a potential difference (voltage) of

$$\begin{aligned}E &= 0.000553 \times 326\\ &= 0.180 \text{ V}\\ &= 180 \text{ mV}\end{aligned}$$

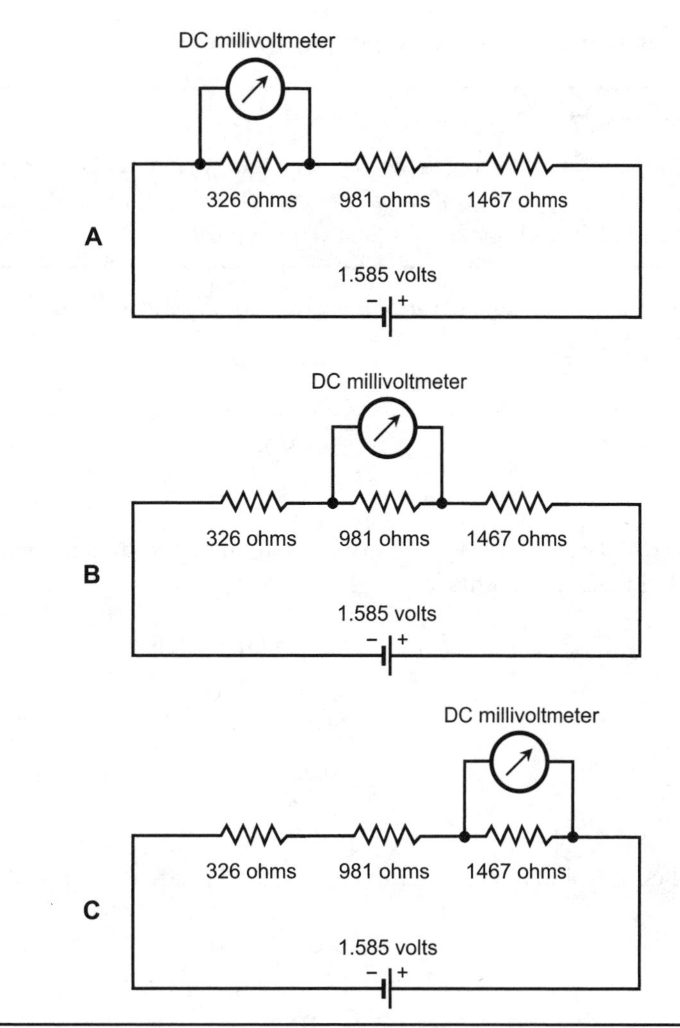

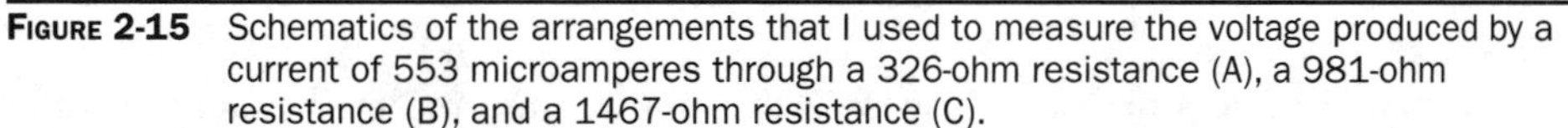

FIGURE 2-15 Schematics of the arrangements that I used to measure the voltage produced by a current of 553 microamperes through a 326-ohm resistance (A), a 981-ohm resistance (B), and a 1467-ohm resistance (C).

When I made the measurement, I got 186 mV, which was 6 mV more than the predicted potential difference. My experimental error was therefore

$$
\begin{aligned}
&[(186 - 180)/180] \times 100 \\
&= (6/180) \times 100 \\
&= 0.0333 \times 100 \\
&= +3.33\% \\
&\approx +3\%
\end{aligned}
$$

The plus sign indicates that my experimental error was positive in this case.

Did You Know?

In science and engineering, a "wavy" or "squiggly" equals sign indicates that a result has been rounded off or approximated. You'll often see this symbol in papers, articles, and books that deal with experiments.

Across the 981-ohm resistor alone (Fig. 2-15B), I predicted that I would see a potential difference of

$$\begin{aligned} E &= 0.000553 \times 981 \\ &= 0.542 \text{ V} \\ &= 542 \text{ mV} \end{aligned}$$

I measured 560 mV, which was 18 mV more than the predicted potential difference. I calculated the experimental error as

$$\begin{aligned} &[(560 - 542)/542] \times 100 \\ &= (18/542) \times 100 \\ &= 0.0332 \times 100 \\ &= +3.32\% \\ &\approx +3\% \end{aligned}$$

Across the 1467-ohm resistor alone (Fig. 2-15C), I predicted a potential difference of

$$\begin{aligned} E &= 0.000553 \times 1467 \\ &= 0.811 \text{ V} \\ &= 811 \text{ mV} \end{aligned}$$

I measured 838 mV, which was 27 mV above the predicted potential difference. Calculating the experimental error, I got

$$\begin{aligned} &[(838 - 811)/811] \times 100 \\ &= (27/811) \times 100 \\ &= 0.0333 \times 100 \\ &= +3.33\% \\ &\approx +3\% \end{aligned}$$

CHAPTER 3

Capacitors

Electrical resistance slows the flow of AC or DC charge carriers (usually electrons) by "brute force." Electrical *capacitance* impedes the flow of AC charge carriers by storing energy in the form of an *electric field*.

What Is Capacitance?

Imagine two gigantic, flat sheets of electrically conducting metal. You place them parallel to each other, and separate them by a few centimeters. If you connect them to the terminals of a battery, as shown in Fig. 3-1, the sheets will charge up, one with positive polarity and the other with negative polarity.

Because the plates are large, it will take some time for the negative one to "fill up" with extra electrons, and it will take an equal amount of time for the positive one to have its electrons "drained out." Eventually, the potential difference between the two plates will equal the battery voltage, and an electric field will fill the space between them. Figure 3-2 is a relative graph of the electric field intensity as a function of time.

Physicists define *capacitance* as the ability of plates such as these, and of the space between them, to store electrical energy. You can abbreviate the word "capacitance" by writing an uppercase, italic letter C. A *capacitor* is a component built specifically to introduce a certain amount of capacitance into electrical circuits.

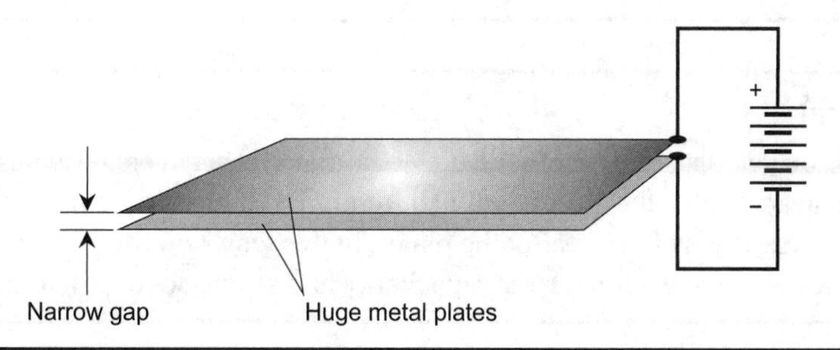

FIGURE 3-1 Two gigantic, parallel, closely spaced conducting plates can store the energy from a battery as an electric field.

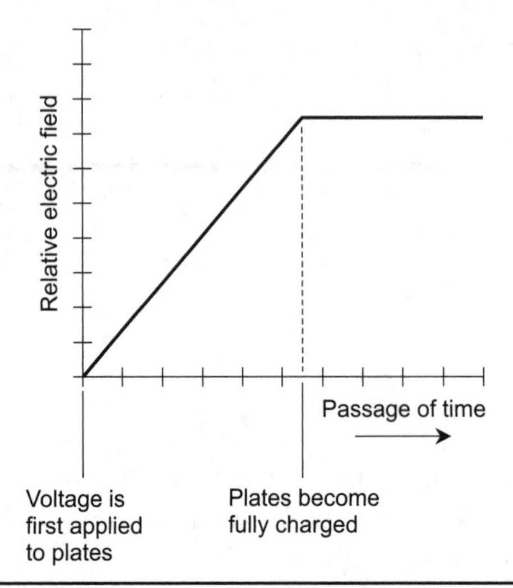

FIGURE 3-2 Relative electric field intensity, as a function of time, between two metal plates connected to a voltage source.

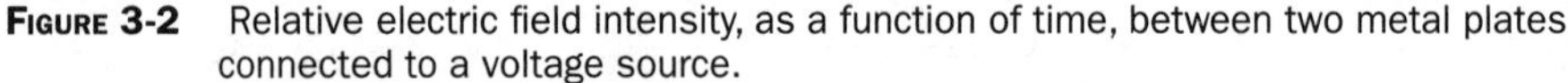

Simple Capacitors

You can place two sheets or strips of thin metal foil together, keeping them evenly separated by a thin layer of nonconducting (also called *dielectric*) material, such as paper or plastic, and then roll up the assembly to obtain a large mutual surface area in a small physical space. When you do this, the electric field quantity and intensity between the plates can get large enough so that the device exhibits considerable capacitance. Alternatively, you can take two separate sets of several plates and mesh them together with air or solid dielectric sheets between them.

Did You Know?

In the early years of electricity as a science in its own right, capacitors were called *condensers* because they could concentrate, or "condense," potential energy in the form of an electric field.

Aha!

When you place a layer of solid dielectric material between the plates of a capacitor (instead of air), the electric field concentration increases without your having to increase the surface areas of the plates. In this way, you can get a physically small component to exhibit a large capacitance and store a lot of potential energy.

The voltage that a capacitor can handle depends on the thickness of the metal sheets or strips, on the spacing between them, and on the type of dielectric material that manufacturers use to build the component. In general, the capacitance varies in direct proportion to the mutual surface area of the conducting plates or sheets, but inversely according to the amount of separation between the conducting sheets.

- If you maintain constant spacing between the sheets and increase their mutual area, the capacitance goes up.
- If you maintain constant spacing between the sheets and decrease their mutual area, the capacitance goes down.
- If you maintain constant mutual area and move the sheets closer together, the capacitance goes up.
- If you maintain constant mutual area and move the sheets farther apart, the capacitance goes down.

Capacitance also depends on the *dielectric constant* of the material between the metal sheets or plates of a capacitor. Scientists assign a dielectric constant of exactly 1 to a vacuum. Poor conductors of electricity almost always have dielectric constants greater than 1—sometimes vastly greater.

If you take a capacitor in which a vacuum exists between the metal sheets or plates, and then fill up all of the space with a material whose dielectric constant equals k, then the capacitance will increase by a factor of k. Table 3-1 lists the dielectric constants for various common substances.

TABLE 3-1 Dielectric constants for media used in capacitor manufacture. Except for air and a vacuum, these values are approximate.

Substance	Dielectric Constant (approx.)
Air, dry, at sea level	1.0
Glass	4.8–8.0
Mica	4.0–6.0
Mylar	2.9–3.1
Paper	3.0–3.5
Plastic, hard, clear	3.0–4.0
Polyethylene	2.2–2.3
Polystyrene	2.4–2.8
Polyvinyl chloride	3.1–3.3
Porcelain	5.3–6.0
Quartz	3.6–4.0
Strontium titanate	300–320
Teflon	2.0–2.2
Titanium oxide	160–180
Vacuum	1.0

Expressing Capacitance

When you connect a battery between the plates of a capacitor, the voltage between the plates builds up at a rate that depends on the capacitance. A capacitance of one *farad* (1 F) represents a current flow of one ampere (1 A) while the voltage increases at the rate of one volt per second (1 V/s).

The farad represents a huge amount of capacitance. You'll almost never see a real-world capacitor with a value of 1 F. Engineers and technicians (and you!) express capacitance values in terms of the *microfarad* (μF) and the *picofarad* (pF), such that

$$\begin{aligned} 1\ \mu\text{F} &= 0.000001\ \text{F} \\ &= 10^{-6}\ \text{F} \end{aligned}$$

and

$$\begin{aligned} 1\ \text{pF} &= 0.000001\ \mu\text{F} \\ &= 10^{-12}\ \text{F} \end{aligned}$$

Commercial vendors can produce physically small components that have fairly large capacitance values. Conversely, some capacitors with small values take up a large physical space. The physical size of a capacitor, if all other factors are held constant, varies in direct proportion to the *voltage* that it can handle before the dielectric material breaks down under the electrical stress. As the rated voltage increases, the component generally grows bigger and more massive.

Fixed Capacitors

A *fixed capacitor* has a value that you can't adjust, and that (ideally) does not vary when environmental or circuit conditions change. Several common types of fixed capacitors have been used for decades.

Paper Capacitors

In the early days of electronics, capacitors were commonly made by placing paper soaked with mineral oil between two strips of foil, rolling the assembly up (Fig. 3-3), attaching wire leads to the two pieces of foil, and enclosing the rolled-up foil and paper in an airtight cylindrical case. You'll find so-called *paper capacitors* in older electronic equipment, such as vintage radios. They have values ranging from about 0.001 μF to 0.1 μF, and can handle up to about 1000 V.

Mica Capacitors

Mica is a naturally occurring, solid, transparent mineral substance that flakes off in thin sheets. It makes an excellent dielectric for capacitors. *Mica capacitors* can be manufactured by alternately stacking metal sheets and layers of mica, or by applying silver ink to sheets of mica. The metal sheets are wired together into two meshed sets, forming the two terminals of the capacitor, as shown in Fig. 3-4.

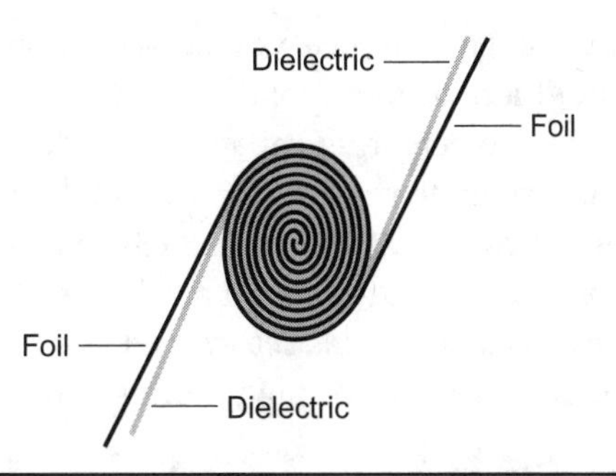

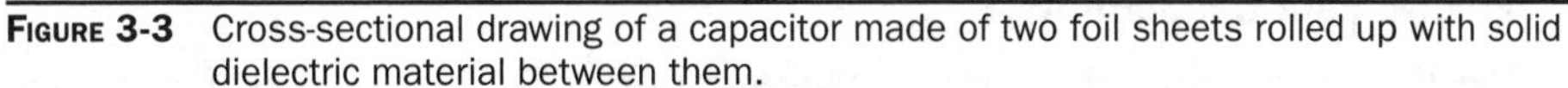
FIGURE 3-3 Cross-sectional drawing of a capacitor made of two foil sheets rolled up with solid dielectric material between them.

Mica capacitors have *low loss* (high efficiency) as long as you don't subject them to excessive voltage, in which case internal arcing can destroy the dielectric. Mica-capacitor voltage ratings can range from a few volts (with thin mica sheets) up to several thousand volts (with thick mica sheets and thick metal plates). Capacitance values range from a few tens of picofarads up to approximately 0.05 μF.

> **Tip**
> Mica capacitors are bulky in proportion to their capacitance, but they work well in wireless receivers and transmitters, and you can expect them to maintain constant capacitance and high efficiency for decades.

Ceramic Capacitors

Ceramic materials perform well as dielectrics. Sheets of metal can be stacked alternately with wafers of ceramic to make *ceramic capacitors* according to the geometry of Fig. 3-4. Ceramic, like mica, has low loss.

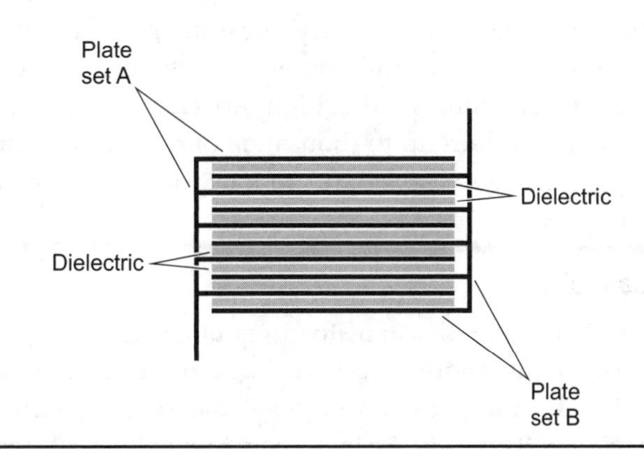

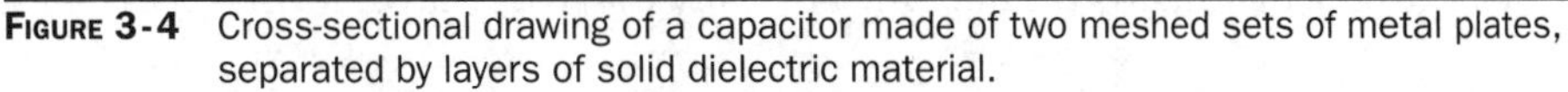
FIGURE 3-4 Cross-sectional drawing of a capacitor made of two meshed sets of metal plates, separated by layers of solid dielectric material.

For small values of capacitance, you need only one disk-shaped layer of ceramic material; you can glue two metal plates to the disk, one on each side, to obtain a *disk-ceramic capacitor*. To get larger capacitance values, you can stack layers of metal and ceramic, connecting alternate layers of metal together as the electrodes.

Another method exists for the manufacture of these capacitors. You can start with a cylinder of ceramic material and apply metal "ink" to its inside and outside surfaces to make a *tubular ceramic capacitor*. They have values ranging from a few picofarads to about 0.5 μF and voltage ratings comparable to paper capacitors.

Plastic-Film Capacitors

Plastics make good dielectrics for the manufacture of capacitors. *Polyethylene* and *polystyrene* are commonly used. The method of manufacture resembles that for paper capacitors. Stacking methods can also work with plastic. The geometries vary, so you'll find *plastic-film capacitors* in various shapes. Capacitance values range from about 50 pF to several tens of microfarads. Most often, you'll encounter values from approximately 0.001 μF to 10 μF.

Plastic capacitors function well in electronic circuits at all frequencies, and at low to moderate voltages. They exhibit good efficiency, although not as high as that for mica-dielectric or air-dielectric capacitors.

Electrolytic Capacitors

All of the above-mentioned types of capacitors provide relatively small values of capacitance. They are *nonpolarized*, meaning that you can hook them up in either direction. In some cases, the vendor offers a recommendation as to which side should go to signal ground. An *electrolytic capacitor*, in contrast, is *polarized*; it provides approximately 1 μF up to thousands of microfarads, but you *must* connect it in the proper direction if you want it to work.

Component manufacturers assemble electrolytic capacitors by rolling up multiple layers of aluminum foil strips separated by paper saturated with an *electrolyte* liquid that conducts current. When DC flows through the component, the aluminum oxidizes because of chemical interaction with the electrolyte. The oxide layer does not conduct and, therefore, forms the dielectric for the capacitor. The layer is extremely thin, yielding high capacitance per unit volume. Electrolytic capacitors can have values up to thousands of microfarads, and some can handle thousands of volts. These capacitors are most often seen in audio amplifiers and DC power supplies.

Tantalum Capacitors

Another type of electrolytic capacitor uses tantalum rather than aluminum. The tantalum capacitor can comprise foil like the aluminum in a conventional electrolytic capacitor, or take the form of a porous pellet, the irregular surface of which provides a large area in a small volume. An extremely thin oxide layer forms on the tantalum.

Tantalum capacitors have high reliability and excellent efficiency. You'll see them in military and aerospace environments (or anywhere else where technicians find servicing inconvenient or impossible) because they almost never fail. Tantalum capacitors have values similar to those of aluminum electrolytics, and they work well in audio and digital circuits as replacements for aluminum types.

Variable Capacitors

You can change the values of certain capacitors by adjusting the mutual surface area between the plates, or by changing the spacing between the plates. Two main types of *variable capacitors* exist: the *air-variable capacitor* and the *trimmer capacitor*. You'll occasionally encounter a less common type known as a *coaxial capacitor*.

Air-Variable Capacitors

You can assemble an air-variable capacitor by connecting two sets of metal plates so that they mesh, and by affixing one set to a rotatable shaft. The rotatable set of plates constitutes the *rotor*, and the fixed set constitutes the *stator* (Fig. 3-5). You'll find air variables in vintage radio receivers (particularly those that used *vacuum tubes* rather than semiconductor components), and in high-power wireless antenna tuning networks.

Air variables have maximum capacitance that depends on the number of plates in each set, and also on the spacing between the plates. Common maximum values range from 50 to 500 pF; occasionally you'll find an air variable that can go up to 1000 pF. Minimum values are a few picofarads. The voltage-handling capability depends on the spacing between the plates. Some air variables can handle several kilovolts (kV) at high AC frequencies.

Air variables can still be seen in older "shortwave radios" designed to work at frequencies from approximately 500 kHz to 30 MHz. Air variables offer high

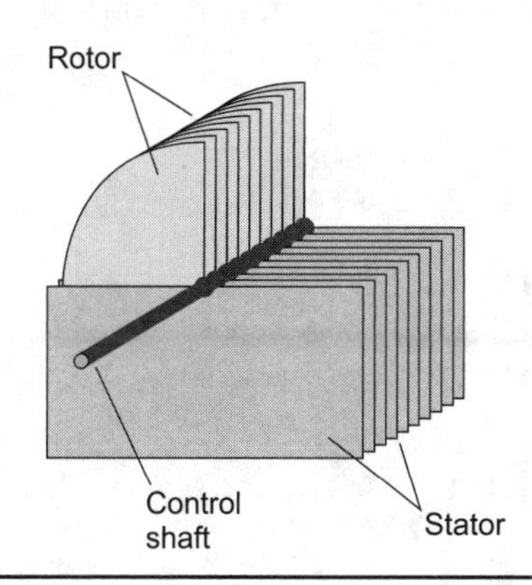

FIGURE 3-5 Simplified drawing of an air-variable capacitor.

efficiency and excellent *thermal stability* (meaning that their values don't appreciably change with wild fluctuations in the ambient temperature).

> **Tip**
>
> Although air variables lack electrical polarization, you'll usually want to connect the rotor plates, along with the control shaft, to the metal chassis or circuit board perimeter, which constitutes the *common ground*. That arrangement will minimize the effects of stray capacitance from your body affecting the component when you touch, or bring your hand near, the rotor shaft to adjust it.

Trimmer Capacitors

When you rarely need to change the value of a capacitor, you can use a *trimmer capacitor* in place of the more expensive, and bulkier, air variable. A trimmer consists of two plates, mounted on a ceramic base and separated by a sheet of solid dielectric. You can vary the spacing between the plates with an adjusting screw, as shown in Fig. 3-6. Some trimmers contain two interleaved sets of multiple plates, alternating with dielectric layers to increase the capacitance.

You can connect a trimmer capacitor in parallel with an air variable, facilitating exact adjustment of the latter component's minimum capacitance. Some air-variable capacitors have trimmers built in to serve this purpose. Typical maximum values for trimmers range from a few picofarads up to about 200 pF. They handle low to moderate voltages, exhibit excellent efficiency, and are nonpolarized.

Coaxial Capacitors

You can use two telescoping sections of metal tubing to build a *coaxial capacitor* (Fig. 3-7). The device works because of the variable effective surface area between the inner and outer tubing sections. A sleeve of plastic dielectric separates the sections of tubing, allowing you to adjust the capacitance by sliding the inner section in or out of the outer section. Coaxial capacitors work well in high-frequency

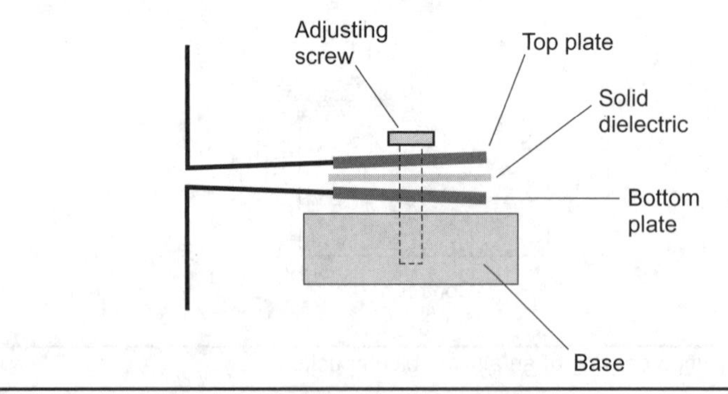

FIGURE 3-6 Cross-sectional drawing of a trimmer capacitor.

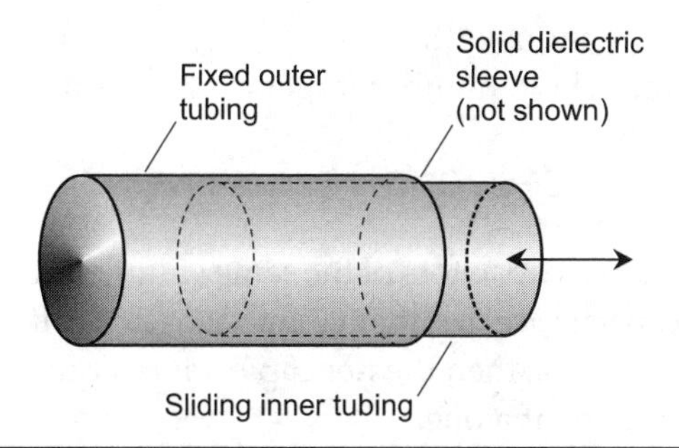

FIGURE 3-7 Simplified drawing of a coaxial variable capacitor.

AC applications, particularly in wireless antenna tuners. Their values range from a few picofarads up to approximately 100 pF.

Stray Capacitance

Any two pieces of conducting material in close proximity can act as a capacitor. Often, such *interelectrode capacitance* is so small (a picofarad or less) that you don't have to worry about it. In utility and *audio-frequency* (AF) circuits, interelectrode capacitance rarely poses trouble, but it can cause problems in *radio-frequency* (RF) systems. The risk of trouble increases as the frequency increases.

The most common consequences of excessive interelectrode capacitance are *feedback* and unwanted changes in the characteristics of a circuit with variations in the operating frequency. You can minimize the interelectrode capacitance in an electronic device or system by keeping the interconnecting wires as short as possible within each individual circuit, by using shielded cables to connect circuits to each other, and by enclosing the most sensitive circuits in metal housings to electrically isolate them.

Handy Math

In series and parallel circuits, capacitance values combine differently than resistors do in similar arrangements. The basic formulas follow, along with an additional formula that allows you to calculate the extent to which a given capacitor will impede the flow of AC.

Capacitors in Series

When you connect two or more capacitors in series, their values combine in the same way that resistances combine in parallel (assuming that no stray capacitance exists). In general, if you connect several capacitors in series, you'll observe a net capacitance smaller than that of any of the individual components. You should always use the same size units when you calculate the net capacitance of any combination.

Consider n capacitors with values C_1, C_2, C_3, ..., C_n connected in series, where n represents some whole number larger than 1. You can find the net capacitance C using the formula

$$C = 1/(1/C_1 + 1/C_2 + 1/C_3 + \ldots + 1/C_n)$$

If you connect two capacitors of the same value in series, then the net capacitance equals half the capacitance of either component alone. If you connect n capacitors of the same value in series, then the net capacitance equals $1/n$ times the capacitance of any single component alone.

If you have only two capacitors in series, one with value C_1 and the other with value C_2 in the same units as C_1, then you can use the formula

$$C = C_1C_2/(C_1 + C_2)$$

Tip

If you connect two or more capacitors in series, and if one of them has a capacitance that's *far smaller* than any of the others, then the net capacitance roughly equals the *smallest* capacitance.

Capacitors in Parallel

Capacitances in parallel add like resistances in series. The net capacitance equals the sum of the individual component values, as long as you use the same units all the way through your calculations.

Suppose that you connect capacitors C_1, C_2, C_3, ..., C_n in parallel. As long as no stray capacitance exists, you can calculate the net capacitance C with the formula

$$C = C_1 + C_2 + C_3 + \ldots + C_n$$

Tip

If you parallel-connect two or more capacitors, and if one of them has a capacitance that's *far greater* than any of the others, then for practical purposes the net capacitance roughly equals the *largest* capacitance.

Capacitive Reactance

If you specify the frequency of an AC source (in hertz) as f, and if you specify the capacitance of a component (in farads) as C, then you can calculate the *capacitive reactance* (in ohms, as a mathematically imaginary quantity) using the formula

$$X_C = -1/(2\pi f C)$$
$$\approx -1/(6.2832 f C)$$

This formula also works if you input f in megahertz and C in microfarads.

Heads Up!

The arithmetic for calculating capacitive reactance can get tricky, and if you don't exercise caution, you can make mistakes. You have to work with reciprocals, so the numbers can get awkward. Also, you have to watch those negative signs. They're critical when you want to draw graphs describing systems containing reactance.

Experiment 1: Discharging a Capacitor

Large-value capacitors can hold a charge for a long time. That's cool, but in some applications, it's also dangerous. For example, this *residual charge* can pose a shock hazard to technicians who test, maintain, or repair a high-voltage DC power supply. If you want to ensure that a big capacitor won't stay charged indefinitely, you must connect a resistor across it to gradually drain the charge after you disconnect the voltage source from it. In this experiment, you'll see how fast resistors can discharge an electrolytic capacitor.

Meter Upgrade

Earlier in this book, I recommended that you get a digital multimeter. If you bought a basic meter, such as the one described in Chapter 1, you'll probably want to get a better one after a while. I did! You don't have to spend a fortune to get a good meter, but you shouldn't scrimp too much either. I chose the Radio Shack True RMS (T-RMS) digital multimeter. Figure 3-8 is a photograph of this instrument.

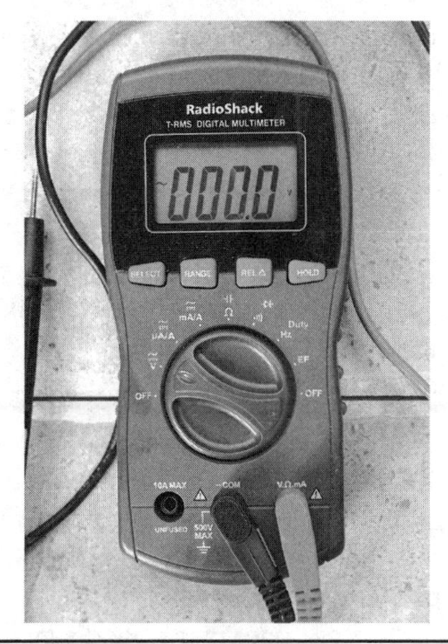

FIGURE 3-8 The Radio Shack T-RMS digital multimeter.

The T-RMS has a rotary switch with fewer positions than the ones on some less sophisticated meters, but don't let that deceive you! Buttons allow you to select from several ranges or functions for each position of the central switch. The T-RMS does share one feature with all other meters of its kind: The main switch has an "OFF" position. Whenever you finish using the meter, use this switch to shut it off. That way, you'll maximize the life of the meter's internal battery.

When you get a meter like the T-RMS, read the instruction manual to make sure that you get the most out of it. You'll get a pleasant surprise when you find out how many things this device can do, but you'll have to read that manual to learn it all. I did, going against my innate fondness for the TECISASWH ("Tweak Every Control In Sight And See What Happens") method of climbing learning curves.

Here's a Twist!

A failure of my original digital meter proved a major motivator in my decision to upgrade. The display on that old meter started showing impossible characters, suggesting that the thing had met an irreversible demise. Sooner or later, all lab test hardware will stop working. Keep that fact in mind when you decide how much you want to spend on a meter or other test instrument! I recommend that you buy a good one, but don't get too extravagant.

Hold That Charge!

Obtain a 2200-μF, 50-V electrolytic capacitor from Radio Shack or an online vendor, and a 6-V lantern battery from your local hardware or department store. Set the meter to read DC voltage. The T-RMS can automatically find its own range without your having to preselect it, so you need only make sure that you set it to read DCV, not ACV. Test your battery to verify that it hasn't weakened or died on the shelf at the store! You should get a reading of a little more than 6 V. My T-RMS meter told me that my battery had plenty of gusto, checking out at 6.31 V.

The T-RMS goes into the capacitance-measuring mode when you set the main switch at the center position with the little capacitor symbol and the ohms symbol (clearly visible in Fig. 3-8), and then hit the "SELECT" button so that the display shows a tiny "nF." The meter will find its own best range once it has something to look for. Now you can test your capacitor to make sure that its value comes reasonably close to the specified capacitance.

Tip

The Radio Shack T-RMS digital multimeter sets itself to measure capacitance in *nanofarads* (nF) by default. A nanofarad (1 nF) equals 0.000000001 farad (10^{-9} F) or 0.001 microfarad (0.001 μF). A nanofarad is also the equivalent of 1000 picofarads (1000 pF). The meter will adjust itself to another range if you test a capacitor whose value would not make good sense if expressed in nanofarads.

Connect the meter's black probe tip to the negative end of the capacitor with a clip lead. Connect the meter's red probe tip to the other end of the capacitor with another clip lead. After a few seconds, the meter should adjust itself to *millifarads* (mF), where 1 mF equals 1000 μF. Wait a little longer, and some numbers should pop up on the display. My display showed 2.235, indicating 2.235 mF or 2235 μF.

Warning! **Whenever you work with electrolytic capacitors, wear gloves and safety glasses. Always make sure that the negative (or minus) side of the capacitor goes to the negative terminal on the battery or other DC power source. Otherwise the capacitor might suffer permanent damage. It could even catch on fire or blow up like a firecracker!**

Make sure that the capacitor is fully discharged by shorting it out with a jumper for a couple of seconds. Pay close attention to the capacitor's polarity. Set your digital multimeter to indicate DC volts again, and connect the capacitor directly across the battery. Measure the voltage across the combination (Fig. 3-9A), using clip leads to secure the meter probe connections to the capacitor's wires. When I did that test, I got the same 6.31 V as I saw when I tested the battery all by itself. No surprise there!

Now disconnect one end of the capacitor from the battery, and watch the meter reading (Fig. 3-9B). It should hold almost constant, dropping only a tiny fraction of a volt every few seconds. Obviously, you'll have to wait a long time to see that capacitor discharge all the way. Does this result surprise you? Keep in mind that 2200 μ F is a lot of capacitance. If the component is in good condition, it should hold

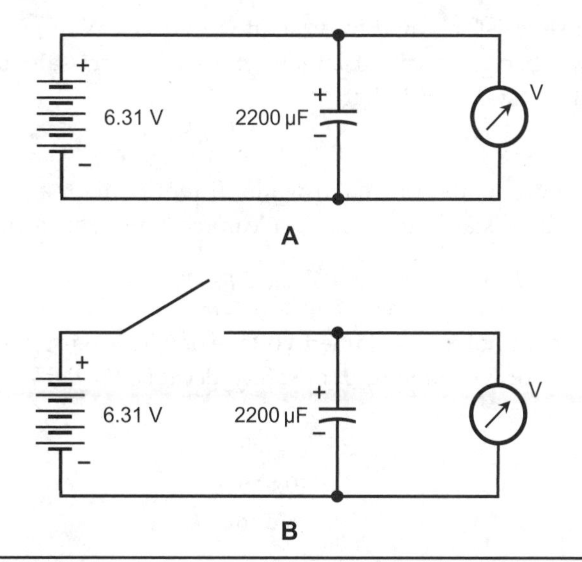

FIGURE 3-9 At A, the capacitor and the meter are both connected across the battery. At B, the meter is still connected across the capacitor, but both are disconnected from the battery.

a charge for hours. Now do you see why, in the olden days of electronics where capacitors held hundreds or even thousands of volts, technicians had to take care around them?

> **Tip**
>
> All voltmeters have high, but not infinite, internal resistance. That resistance will cause the capacitor to discharge more quickly than it would do if the meter weren't connected to it. The T-RMS meter has extremely high resistance in the voltmeter mode, so it doesn't "load down" the capacitor very much. Cheaper meters have lower internal resistance when you use them to measure voltage. You might want to repeat this experiment with your analog meter and see if the capacitor discharges faster through it than it does through the T-RMS. If you have a lot of patience, disconnect the capacitor from everything after fully charging it, let it sit for 24 hours, and then check it again to see how much voltage remains.

Choosing a Resistance

If you connect a resistor across a capacitor, you can force the capacitor to discharge at a rate that depends on the values of the capacitor and the resistor. The larger the value of either component, the longer the capacitor will hold the charge. When you connect a resistor across a capacitor to discharge that capacitor, you must ensure that the resistor has a large enough ohmic value and a high enough power rating to keep it from burning out, but not so large that the capacitor takes too long to lose its charge.

If you use resistors rated to dissipate up to 1/2 W of power continuously and indefinitely, then you can calculate the minimum safe ohmic value for the discharging resistor. Recall that

$$P = E^2/R$$

where P represents the power in watts, E represents the voltage in volts, and R represents the resistance in ohms. You can rearrange this formula to get

$$R = E^2/P$$

My capacitor holds a sustained voltage of 6.31 V with power applied. Let's say that $E = 7$ V to provide a margin of safety. Because $P = 0.5$ W, the calculation works out as

$$\begin{aligned} R &= 7^2/0.5 \\ &= 49/0.5 \\ &= 98 \text{ ohms} \end{aligned}$$

The smallest resistance that you can use without overheating the component is therefore 98 ohms, which you can round up to 100 ohms.

Get some 1/2-W resistors from a vendor, such as Radio Shack. I recommend 3300 (3.3k) ohms and 47,000 (47k) ohms. You can connect either of these resistors directly across your lantern battery and capacitor, and remain comfortably within the safety margin for power dissipation in this circuit. Figure 3-10 is a photograph of all the components that you'll need for this experiment.

Use your ohmmeter to check the resistors to guarantee that they actually have the values that the manufacturer has specified. When I tested my units, they came within 5% of the rated values.

Heads Up!

I had to remind myself to set the ohms to zero with a dead short between the test leads before I tested the resistors with my analog multimeter, reworking that "0Ω ADJ" control every time I changed resistance scales on the meter. Otherwise, I might have thought that a good component was defective!

Tip

If you can't find a 2200-μF capacitor, then you can substitute one with a different value, as long as it's in the same general range. For example, a 1000-μF unit or a 4700-μF unit will work okay. Make sure, however, that the component is rated to handle a good deal more than 6 V.

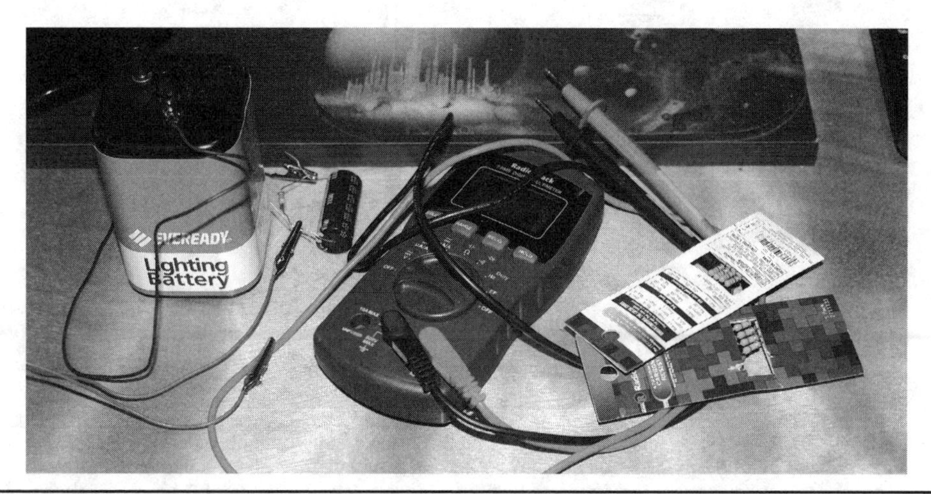

FIGURE 3-10 Components for the discharge experiment include a 6-V battery, a multimeter, two clip leads, a 2200-μF 50-V capacitor, a 3.3k resistor, and a 47k resistor. (Radio Shack sells resistors in packs of five.)

Bleed Off the Charge

Discharge the capacitor by shorting it out with a jumper for a couple of seconds. Then connect the 3.3k resistor across the capacitor by twisting the leads of the two components together. Set your meter to indicate DC volts. Connect the meter probe tips to the resistor/capacitor (RC) combination by using clip leads. Connect the other ends of the clip leads to the battery, once again paying close attention to the capacitor polarity (Fig. 3-11A). Note the voltage. It should be the same as it was without the resistor there, assuming that your battery hasn't grown feeble.

Now disconnect the battery from the RC combination but leave the meter connected, as shown in Fig. 3-11B. What do you observe? I saw the voltage decline rapidly, taking only a few seconds to drop below 1 V and then, gradually, draining down toward zero. The discharge process went too fast for me to tabulate values against time, but the action of the resistor was obvious. How long does it take for the voltage to get down to 500 mV? How about 100 mV, or 10 mV, or 1 mV?

Discharge Decrement

Now replace the 3.3k resistor with a 47k resistor. Find a clock or wristwatch with a moving second hand (or a digital seconds display) that's easy to read. Connect the battery to the combination, as shown in Fig. 3-11A. Measure the voltage; it should

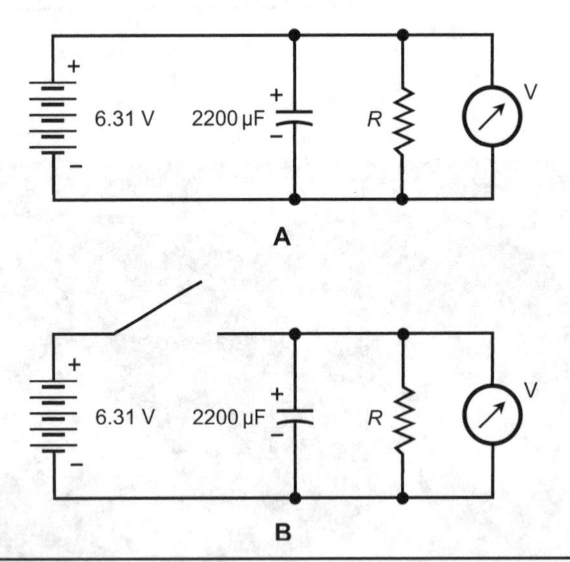

FIGURE 3-11 At A, the capacitor, resistor, and meter are all connected across the battery. At B, the meter is still connected across the capacitor and the resistor, but all three are disconnected from the battery. You can vary the value of R as you wish.

be the same as it was with no resistor at all, or with the 3.3k resistor. Now disconnect the battery from the RC combination, but leave your meter connected across that combination, as shown in Fig. 3-11B. Check and write down the voltage after 10 seconds have gone by, then 20 seconds, then 30 seconds, then 40, 50, 60, and so on. Keep up this exercise for at least 4 minutes at 10- or 15-second intervals. Table 3-2 shows my results.

Now plot the data from your voltage-versus-time table in the form of a graph. Make the horizontal axis show the time after power-down. Make the vertical axis show the voltage across the RC combination. Figure 3-12 shows my graph, based on the data from Table 3-2. Small open circles represent measured data points. The smooth curve approximates the *discharge decrement* for the RC combination. You can expect to get results close to these. You can try other resistors and make tables and graphs for them, too. I suggest values such as 10k, 22k, and 33k.

TABLE 3-2 Measured voltages across a 47k resistor in parallel with a 2200-μF capacitor, as a function of the time, after disconnecting a 6.3-V lantern battery.

Time after Power-Down (Min:Sec)	Voltage across RC Circuit (V)
0:10	5.749
0:20	5.299
0:30	4.815
0:40	4.405
0:50	4.021
1:00	3.670
1:15	3.206
1:30	2.795
1:45	2.423
2:00	2.126
2:15	1.847
2:30	1.625
2:45	1.415
3:00	1.240
3:15	1.086
3:30	0.947
3:45	0.829
4:00	0.725

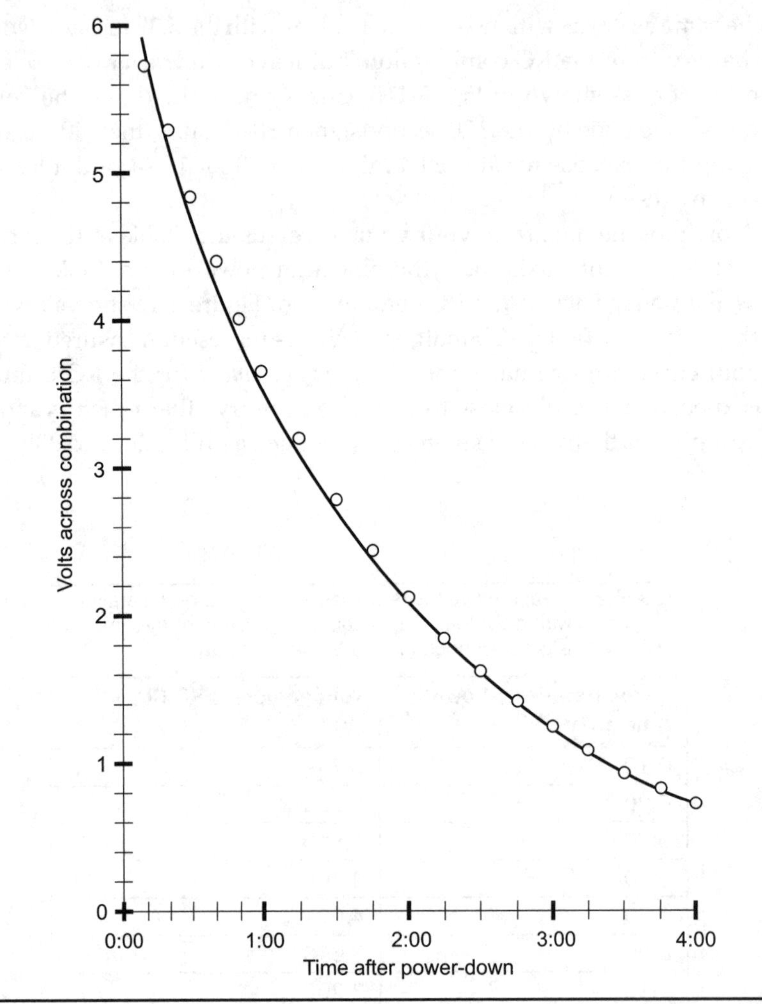

Figure 3-12 Graph of capacitor discharge, showing voltage as a function of time according to the data from Table 3-2.

Experiment 2: Capacitance Measurement

At my local Radio Shack retail store, I found a pack of two capacitors specified as 0.01 μF at 500 V. When I got them home and tested them, I measured one of them as 9.427 nF (0.009427 μF) and the other as 9.769 nF (0.009769 μF). Your values will doubtless differ slightly from mine, so make sure that you test your capacitors individually before connecting them in series or parallel.

Tip

When measuring capacitances less than about 1 μF with a meter such as the T-RMS, use clip leads so that you don't have to touch the wires while the meter is connected to the capacitor. Allow a minute or two for the meter to stabilize. It has to charge up the capacitor to determine its value, and this process can take a while.

Capacitors in Series

Connect the capacitors in series by twisting their wires together. Then connect the T-RMS to the two far-end capacitor wires using clip leads, and let that meter grind out its task! When I carried out this experiment, I got 4.823 nF, which equals 0.004823 μF. Using the quick formula for series capacitances, I would have expected to see, in nanofarads,

$$\begin{aligned} C &= 9.427 \times 9.769 / (9.427 + 9.769) \\ &= 92.09 / 19.20 \\ &= 4.796 \text{ nF} \end{aligned}$$

My result came out high by 0.56%.

Capacitors in Parallel

Take the two capacitors apart by untwisting their wires, taking care not to break either of them off! Then reconfigure the combination so that the two components are connected in parallel. Hook up the meter again and wait the requisite time for it to figure things out. When I did this test, I got 19.20 nF, which corresponds to 0.01920 μF. Using the formula for parallel capacitances, I would have expected to see, in nanofarads,

$$\begin{aligned} C &= 9.427 + 9.769 \\ &= 19.20 \text{ nF} \end{aligned}$$

My result came out exactly on the mark, at least to within the digital resolution of the Radio Shack T-RMS digital multimeter.

CHAPTER 4
Inductors

In this chapter, you'll learn about components that oppose the flow of AC by storing energy as magnetic fields. Then you'll get a chance to build a couple of devices that take advantage of them. Scientists and engineers call them *inductors* because their action is technically known as *inductance*.

What Is Inductance?

Imagine a wire 1,000,000 miles (1,600,000 kilometers) long, roughly the equivalent of two round trips to the moon and back. You straighten the wire out and then make it into a loop, connecting its ends to the terminals of a battery (Fig. 4-1) so that an electric current flows. Because the wire covers such a great span, the electrons need a while to work their way from the negative battery terminal, around the loop, and back into the positive terminal as they "hop" from atom to atom. It will, therefore, take some time for the current to rise to a maximum and stabilize.

The magnetic field produced by the loop will start out small, building up as the electrons make the journey around. Once the electrons reach the positive battery terminal so that a steady current flows around the entire loop, the overall magnetic field quantity (or *flux*) in and near the loop will level off (Fig. 4-2). That field will store energy. The amount of energy that a loop of this sort can store depends on its

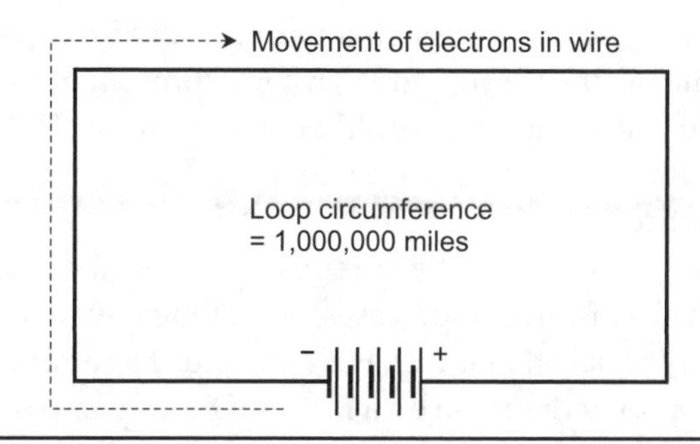

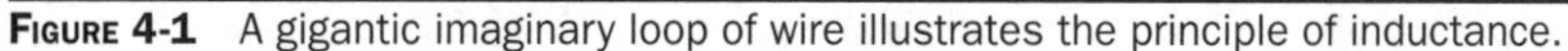

FIGURE 4-1 A gigantic imaginary loop of wire illustrates the principle of inductance.

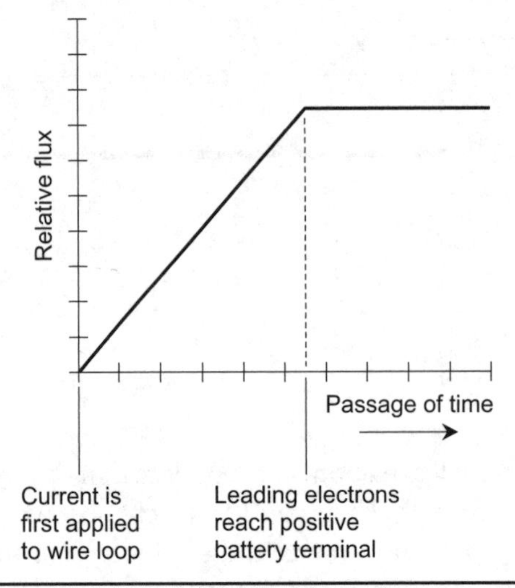

Figure 4-2 Relative magnetic field quantity (or flux) in and around a huge loop of wire connected to a current source, as a function of time.

inductance. Engineers symbolize inductance as a variable quantity by writing an italic, uppercase letter *L*. To abbreviate the word "inductor," write an uppercase, non-italic L.

Simple Inductors

Obviously, you can't make a wire loop 1,000,000 miles in circumference. But you can wind wire into compact coils. When you do that, the magnetic flux for a given length of wire increases compared with the flux produced by a single-turn loop, increasing the inductance. If you place a *ferromagnetic* (or "magnetizable") rod, called a *core* inside a coil, you raise the inductance even more because ferromagnetic materials have high *permeability*, an expression of how much a material concentrates a magnetic field.

By convention, scientists assign a permeability value of 1 to a vacuum. If you have a coil with an air core and you drive DC through the wire, then the flux inside the coil is about the same as it would be in a vacuum. Therefore, the permeability of air equals almost exactly 1. (Actually it's a bit higher, but the difference rarely matters in practice.)

If you place a ferromagnetic core inside a coil, the *flux density* (concentration of magnetic "lines of force") increases, sometimes by a large factor. By definition, the permeability equals that factor. For example, if a certain material causes the flux density inside a coil to increase by a factor of 60, as compared with the flux density in air or a vacuum, then that material has a permeability of 60. Table 4-1 lists the permeability values for a few common substances.

Tip

Plastic and dry wood have properties, in terms of magnetic-field storage, that don't differ much from air or a vacuum; engineers sometimes use these materials as coil cores or "forms" to add structural rigidity to the assembly without significantly changing the inductance. You can, for example, wind some wire on a wooden dowel or hard plastic rod, and it will act like an air-core coil of the same size.

If you hold all other factors constant, the inductance of a helical (spring-shaped) coil, also known as a *solenoid*, increases in direct proportion to the number of turns of wire. The inductance also increases in direct proportion to the diameter of the coil. If you stretch out a coil having a certain number of turns and a certain diameter, while holding all other parameters constant, its inductance goes down. Conversely, if you squash down an elongated coil while holding all other factors constant, the inductance goes up.

TABLE 4-1 Permeability Values for a Few Common Materials

Substance	Permeability (approx.)
Air, dry, at sea level	1.0
Alloys, ferromagnetic	3000–1,000,000
Aluminum	Slightly more than 1
Bismuth	Slightly less than 1
Cobalt	60–70
Iron, powdered and pressed	100–3000
Iron, solid, refined	3000–8000
Iron, solid, unrefined	60–100
Nickel	50–60
Silver	Slightly less than 1
Steel	300–600
Vacuum	1.0 (exact, by definition)
Wax	Slightly less than 1
Wood, dry	Slightly less than 1

Expressing Inductance

When you connect a battery across an inductor, the current builds up at a rate that depends on the inductance. For a given battery voltage, the greater the inductance, the slower the current buildup. The unit of inductance quantifies the ratio between the rate of current buildup and the voltage across an inductor. An inductance of one *henry* (1 H) represents one volt (1 V) across an inductor within which the current increases or decreases at the rate of one ampere per second (1 A/s).

The henry is a huge unit of inductance. You won't often see an inductor this large, although some power-supply filter chokes have inductances up to several henrys. Usually, engineers and technicians express inductance in *millihenrys* (mH), *microhenrys* (μH), or *nanohenrys* (nH). The units relate as follows:

$$\begin{aligned} 1 \text{ mH} &= 0.001 \text{ H} \\ &= 10^{-3} \text{ H} \end{aligned}$$

$$\begin{aligned} 1 \text{ μH} &= 0.001 \text{ mH} \\ &= 10^{-6} \text{ H} \end{aligned}$$

$$\begin{aligned} 1 \text{ nH} &= 0.001 \text{ μH} \\ &= 10^{-9} \text{ H} \end{aligned}$$

Small coils with only a few turns of wire have small inductance, in which the current changes quickly and the induced voltages are small. Large coils with ferromagnetic cores, and having many turns of wire, have high inductance in which the current changes slowly and the induced voltages are large.

Warning! **The current from a battery, building up or dying down through a high-inductance coil, can give rise to a considerable voltage between the end terminals of the coil, sometimes far greater than the voltage of the battery itself. *Spark coils*, such as those used in internal combustion engines, take advantage of this principle. Large coils can present a danger to people ignorant of the wiles of inductance!**

Coil Interaction

In real-world circuits, you'll observe *mutual inductance* between solenoids in close proximity to one another because the magnetic fields extend outside the coils.

Coefficient of Coupling

The *coefficient of coupling*, symbolized k, quantifies the extent to which two inductors interact. Engineers specify k as a number ranging from 0 (no interaction) to 1 (the maximum possible interaction). Two coils separated by a sheet of iron to block magnetic fields, or by a great distance so that the fields are too weak to matter, have a coefficient of coupling of zero ($k = 0$). Two coils wound on the same form, one right over the other, exhibit the maximum possible coefficient of coupling ($k = 1$).

Tip

Some engineers multiply k by 100 and add a percent-symbol (%) to express the coefficient of coupling as a percentage, defining the range $k_{\%} = 0\%$ to $k_{\%} = 100\%$.

Mutual Inductance

You can symbolize the *mutual inductance* between two inductors by writing an uppercase italic M and expressing the quantity in henrys, millihenrys, microhenrys, or nanohenrys. When you have two inductors with values of L_1 and L_2 (both expressed in the same size units) and with a coefficient of coupling equal to k, you can calculate the mutual inductance as

$$M = k\,(L_1 L_2)^{1/2}$$

where the 1/2 power represents the positive square root. When you work out a result using this formula, you'll get the value of M in the same unit as you express L_1 and L_2 (henrys, millihenrys, microhenrys, or nanohenrys).

Effects of Mutual Inductance

If you connect two inductors in series and observe *reinforcing* mutual inductance, you can calculate the total inductance L with the formula

$$L = L_1 + L_2 + 2M$$

where L_1 and L_2 represent the coil inductances and M represents the mutual inductance, all in the same size units. If you connect two coils in series and observe *opposing* mutual inductance, you can calculate the total inductance L with the formula

$$L = L_1 + L_2 - 2M$$

Example 1

Imagine that you connect two coils, having inductances of 30 μH and 50 μH, in series, as shown in Fig. 4-3, so that their fields reinforce. Suppose that the coefficient of coupling equals 0.500. To find the net inductance, you must derive M from k. According to the formula for this purpose,

$$\begin{aligned} M &= 0.500\,(30 \times 50)^{1/2} \\ &= 19.36\ \mu\text{H} \end{aligned}$$

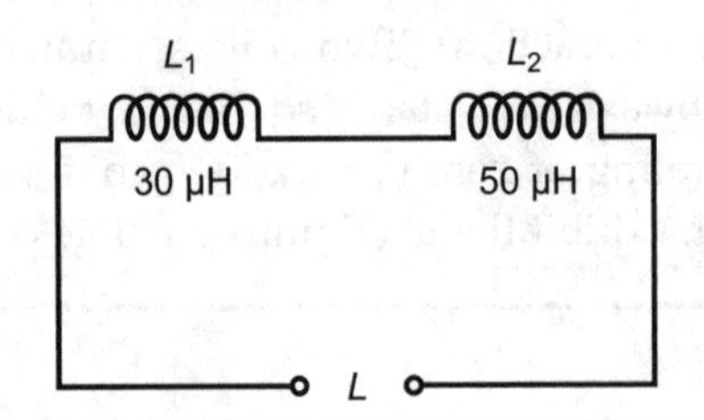

FIGURE 4-3 Illustration for Example 1.

Now you can calculate the total inductance, getting

$$\begin{aligned} L &= L_1 + L_2 + 2M \\ &= 30 + 50 + (2 \times 19.36) \\ &= 118.72\ \mu\text{H} \end{aligned}$$

which you can round off to 120 μH.

Example 2

Imagine two coils with inductances of L_1 = 835 μH and L_2 = 2.44 mH. You connect them in series but in opposite directions (Fig. 4-4), so that their magnetic fields oppose or "buck" each other with a coefficient of coupling equal to 0.922. To find the net inductance, you must convert both values to the same units, preferably microhenrys. Then L_1 = 835 μH and L_2 = 2440 μH, and you can calculate M from k, obtaining

$$\begin{aligned} M &= 0.922\ (835 \times 2440)^{1/2} \\ &= 1316\ \mu\text{H} \end{aligned}$$

Finally, you work out the net inductance as

$$\begin{aligned} L &= L_1 + L_2 - 2M \\ &= 835 + 2440 - (2 \times 1316) \\ &= 643\ \mu\text{H} \end{aligned}$$

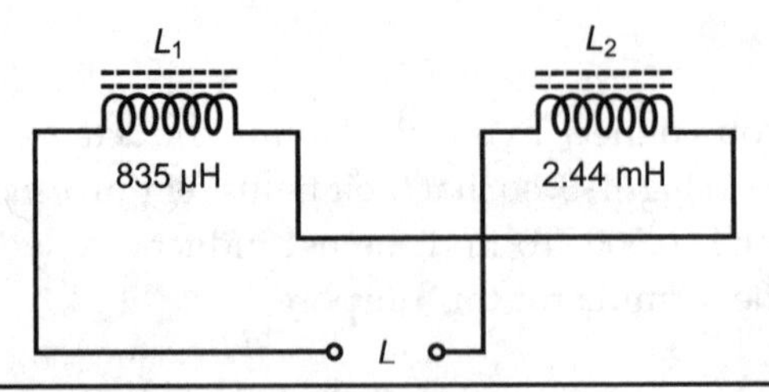

FIGURE 4-4 Illustration for Example 2.

Air Cores

You can wind a coil on a hollow cylinder made of plastic or other non-ferromagnetic material, forming an *air-core coil*. In practice, the attainable inductance for such coils can range from a few nanohenrys up to about 1 mH. The frequency of an applied AC signal does not affect the inductance of an air-core coil, but as the AC frequency increases, smaller and smaller values of inductance produce significant effects.

An air-core coil made of heavy-gauge wire, and having a large radius, can carry high current and can handle high voltage. Air dissipates almost no energy as heat, so air makes an efficient core material even though it has low permeability. For these reasons, air-core coil designs represent an excellent choice for the engineer who wants to build high-power RF transmitters, amplifiers, or tuning networks.

Tip

Air-core coils take up a lot of physical space in proportion to the inductance, especially when designed to withstand high current or high voltage. If you want to wind a coil with a reasonable number of turns, which has an inductance of more than about 1 mH, you'll need a core with higher permeability than air has.

Ferromagnetic Cores

Inductor manufacturers crush samples of ferromagnetic material into dust and then bind the powder into various shapes, providing cores that can greatly increase the inductance of a coil having a given number of turns, as compared with air cores.

Solenoids

Most ferromagnetic-core solenoids have fixed values, but the inductance can be varied somewhat by moving the core in and out of the coil winding, a common practice in old shortwave radio receivers. You can adjust the frequency of a radio's tuned circuit in this way. Because moving the core in and out changes the effective permeability within a coil of wire, this method of frequency adjustment is called *permeability tuning*.

The in/out motion of the ferromagnetic core can be precisely controlled by attaching the core to a screw shaft, and anchoring a nut at one end of a solid plastic cylinder on which the coil is wound. As you turn the screw shaft clockwise, the core enters the cylinder, so the inductance increases. As you turn the shaft counterclockwise, the core moves out of the cylinder, so the inductance decreases.

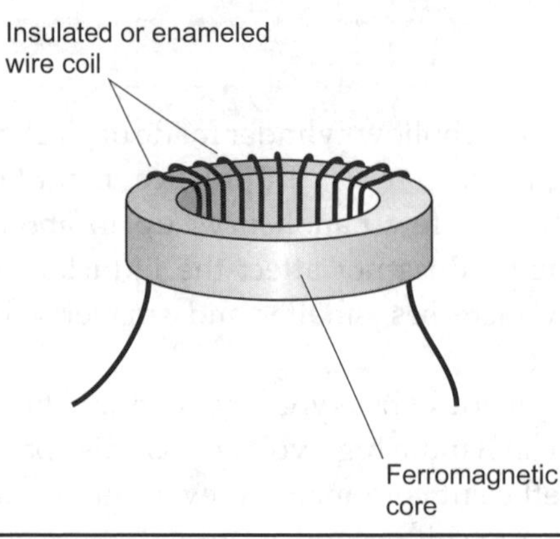

FIGURE 4-5 You can wind a coil around a toroidal ferromagnetic core.

Toroids

These days, if you encounter a coil wound around a ferromagnetic sample, that sample will most likely comprise a *toroid core*, whose shape resembles that of a donut. Figure 4-5 illustrates how you can wind insulated or enameled wire on a ferromagnetic toroid core to make an inductor. Figure 4-6 is a photograph of a toroidal coil on my "old-timer's" wooden breadboard with some other components. (The squares on the breadboard measure 1 inch by 1 inch, so you can get an idea of the coil size.) In addition to physical ruggedness, toroidal coils offer at least three advantages over solenoidal ones:

FIGURE 4-6 A complete toroidal coil on a primitive breadboard in my workshop. The squares on the board measure 1 inch by 1 inch.

1. You'll need fewer turns of wire to get a certain inductance with a toroid, as compared with a solenoid.
2. You can make a toroid physically smaller for a given inductance and current-carrying capacity, as compared with a solenoid.
3. All of the magnetic flux in a toroid remains within the core, eliminating mutual inductance between the coil and other components near it (assuming that you don't want mutual inductance to exist).

On the downside, you'll find toroidal coils more difficult and tedious to wind than solenoidal ones, especially when the winding needs a lot of turns.

Tip

Various vendors sell toroidal cores for coil winding purposes. Perform a search using an Internet search engine on the phrase "toroid cores." Palomar Engineers offers a good selection. At the time of this writing, I found them at **palomar-engineers.com**. They provide instructions with each one of their cores telling you how many turns you'll need to get a particular value of inductance.

Pot Cores

An alternative exists to the toroidal geometry for confining magnetic flux. You can surround a loop-shaped coil of wire with a ferromagnetic shell, as shown in Fig. 4-7, obtaining a *pot core*. A typical pot core has two halves that you clamp around the coil when you've finished winding it. The coil's end leads emerge through small slots in the core.

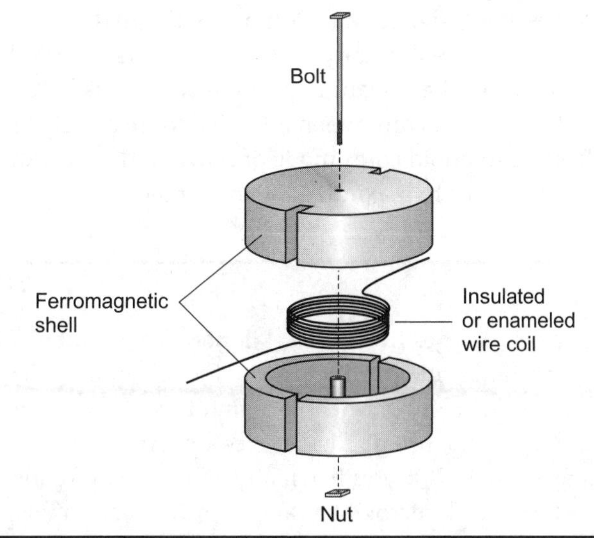

FIGURE 4-7 Exploded view of a pot core. You wind the coil, and then clamp the two parts of the shell together around it.

Pot cores have advantages similar to those of toroids. The shell prevents the magnetic flux from extending outside the physical assembly. You can get far more inductance with a pot core than with a solenoidal coil of comparable physical size. In fact, pot cores work better than toroids if you need a large inductance in a small space.

Tip

Pot-core coils can prove useful over the full audio-frequency (AF) range, and even below the range of human hearing (less than 20 Hz)! However, pot-core coils don't function very well at frequencies above a few hundred kilohertz. They simply have too much inductance for high-frequency use.

Core Saturation

If a ferromagnetic-core coil carries more than a certain amount of current, the core will *saturate*. This phenomenon can occur with any core geometry—solenoid, toroid, or pot core. When an inductor core operates in saturation, it holds as much magnetic flux as it possibly can. Any further increase in the coil current will fail to produce an increase in the core's magnetic flux. In extreme cases, saturation can cause a coil to waste considerable power as heat. That's called *core loss*.

Chokes

A *choke* is an inductor that passes DC, while blocking AC above a certain frequency range. Typically, an inductor offers no little or no opposition to DC (a frequency of zero), and increasing opposition to AC as the frequency goes up.

Some chokes are designed to cut off only RF signals; others block AF as well as RF. Some chokes cut off essentially all AC, even 50-Hz or 60-Hz utility current, and find use in power supplies meant to provide battery-like DC from an AC source.

Small chokes have air cores when intended for relatively high radio frequencies. Larger chokes use solenoid cores made of powdered iron. Other chokes are wound on toroidal or pot cores to maximize the inductance.

Tech Tidbit

The RF spectrum ranges from a few kilohertz to hundreds of gigahertz. At the low end of this range, most inductors use ferromagnetic cores. As the frequency increases, cores with low permeability find favor. You'll often see toroids in RF systems designed for use at frequencies up through about 30 MHz. At very high frequencies (VHF), you'll usually find air-core coils. In the ultra high frequency (UHF) and microwave ranges, inductors can take exotic forms, such as lengths of transmission line or foil runs on printed circuit boards!

Handy Math

In series and parallel circuits, inductances combine in the same way as resistances do. The basic formulas follow, along with a formula, which you can use to calculate the extent to which an inductor impedes (opposes) AC, another formula for finding the approximate inductance of an air-core coil with known dimensions, and yet another formula for finding the *resonant* (natural) frequency of a circuit containing an inductor and a capacitor.

Inductances in Series

Inductances in series add together like resistances in series. The net inductance equals the sum of the individual component values, as long as you use the same units (such as millihenrys, microhenrys, or nanohenrys) all the way through your calculations.

Suppose that you connect two or more inductors L_1, L_2, L_3, ..., L_n in series. Provided that no mutual inductance exists between or among the coils, you can calculate the net inductance L as

$$L = L_1 + L_2 + L_3 + \ldots + L_n$$

> **Tip**
>
> If you series-connect two or more inductors, and if one of them has a value that's *far greater* than any of the others, then for most practical purposes the net inductance roughly equals the *largest* value.

Inductors in Parallel

When you connect multiple inductors in parallel, their values combine in the same way that resistances combine in parallel, assuming that no mutual inductance exists. In general, if you connect several inductors in parallel, you'll observe a net inductance smaller than that of any of the individual components. You should, of course, use the same size units for each of those components.

Consider n inductors with values L_1, L_2, L_3, ..., L_n connected in parallel, where n represents a whole number of 2 or more. You can find the net inductance L using the formula

$$L = 1/(1/L_1 + 1/L_2 + 1/L_3 + \ldots + 1/L_n)$$

If you connect two inductors having identical values in parallel, then the net inductance equals half the inductance of either component alone. If you connect n inductors of the same value in parallel, then the net inductance equals $1/n$ times the inductance of any single component alone.

If you have only two inductors in parallel, one with value L_1 and the other with value L_2 in the same units as L_1, then you can work out the net inductance L as

$$L = L_1 L_2 / (L_1 + L_2)$$

> **Tip**
>
> If you connect two or more inductors in parallel, and if one of them has a value that's *far smaller* than any of the others, then for most practical purposes the net inductance roughly equals the *smallest* value.

Inductive Reactance

If you specify the frequency of an AC signal or current (in hertz) as f, and you specify the inductance of a component (in henrys) as L, then you can calculate the inductive reactance (in ohms, as a mathematically imaginary quantity) using the formula

$$X_L = 2\pi f L$$
$$\approx 6.2832 f L$$

This formula also works if you input f in megahertz and L in microhenrys.

Reactance is a measure of the extent to which a component having inductance and/or capacitance opposes AC. (Resistance, in contrast, involves DC but not AC.)

Air-Core Solenoid Winding Formula

If you want to wind your own air-core inductor (usually around a plastic or wood form to give it some mechanical strength), you can get a good idea of the inductance it will have, provided that you wind it in a single layer and keep the spacing between all the windings constant.

Let's say that your coil measures s inches from end to end and has a radius (not diameter) of r inches. You wind n turns around the form, which constitutes a cylinder or solid plastic or wood rod having the same radius all along its length. As long as you don't give the coil ridiculous dimensions (for example, a 2-foot-radius multiturn loop wound inside a piece of garden hose, or a 36-inch-long coil wound on a wooden dowel only 0.5 inches in radius), then you can calculate the inductance L in microhenrys as approximately

$$L = (r^2 n^2)/(9r + 10s)$$

Figure 4-8 shows the geometry along with the formula and the parameters. If you want to use centimeters rather than inches, convert all dimensions to inches and then calculate. Multiply centimeters by 0.394 to get inches.

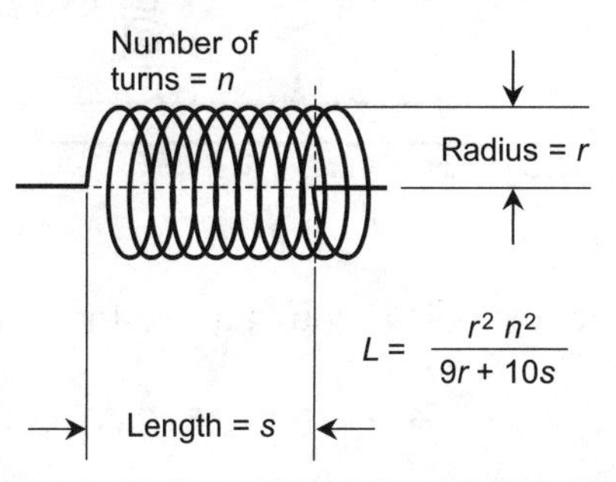

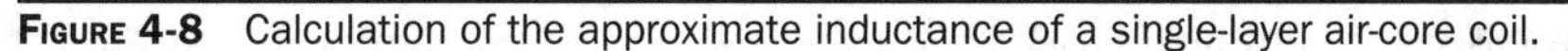

FIGURE 4-8 Calculation of the approximate inductance of a single-layer air-core coil.

Resonance

In a series or parallel resistance-inductance-capacitance (*RLC*) connection, the inductive reactance cancels out the capacitive reactance at a single, well-defined resonant (natural) frequency. This condition is called *resonance*.

> **Tech Tidbit**
>
> Mathematically, inductive reactance has positive-imaginary-number values, while capacitive reactance has negative-imaginary-number values. Resistors have no reactance, and resistance values are expressed as real numbers.

When the inductive and capacitive reactances have equal magnitude (but opposite mathematical signs), the net reactance equals zero. Then only the resistance R remains, opposing the flow of AC, but having no effect on the resonant frequency. Figure 4-9 shows a series *RLC* circuit. Figure 4-10 shows a parallel *RLC* circuit.

You can calculate the resonant frequency f_o of either a series or parallel *RLC* circuit in terms of the inductance L in henrys and the capacitance C in farads, using the formula

$$f_o = 1 / [2\pi(LC)^{1/2}]$$

R C L

FIGURE 4-9 A series resistance-inductance-capacitance (*RLC*) circuit.

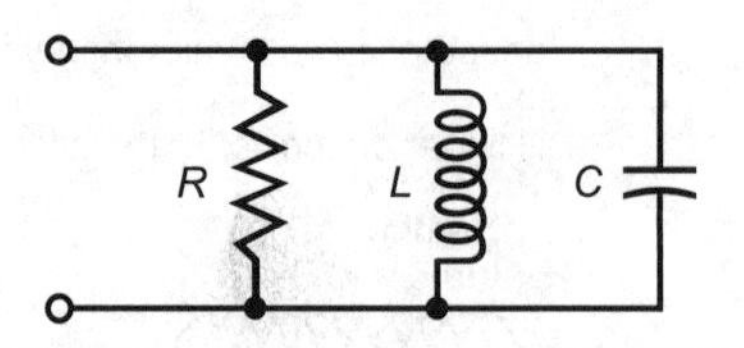

Figure 4-10 A parallel *RLC* circuit.

Because π equals about 3.1416, you can use that number and a little algebra to simplify, getting

$$f_o = 0.15915/(LC)^{1/2}$$

The formula will also work if you want to find f_o in megahertz, and you know *L* in microhenrys and *C* in microfarads.

> **Tip**
> Communications engineers and technicians take advantage of *RLC* resonance in RF transmitting systems in which the physical components have reasonable size in moderate- and high-power applications. An example is the output circuit of an RF power amplifier for use by ham radio operators.

Experiment 1: Electromagnet

In this experiment, you'll build a device to demonstrate the fact that a current-carrying inductor produces a magnetic field. You'll need a 1/2-inch-diameter, 12-inch-long threaded steel rod, two nuts to fit the rod, 20 feet of two-wire lamp cord, some electrical tape, a fresh 6-V lantern battery, a magnetic compass, and a pair of safety glasses.

Theory

When you place a ferromagnetic rod inside a solenoidal coil of wire and then connect the coil to a source of current, you get an electromagnet (Fig. 4-11). The magnetic field produced by the current (provided that it's DC) temporarily magnetizes the core. The core acts like a conventional bar magnet as long as you keep current flowing in the coil.

The strength of an electromagnet depends on the current in the coil, the number of coil turns, and the permeability of the core. Unless the core has been premagnetized, an electromagnet produces a significant magnetic field only when the coil carries current. When the current stops, the magnetic field collapses, although a tiny amount of *residual magnetism* might remain in the core.

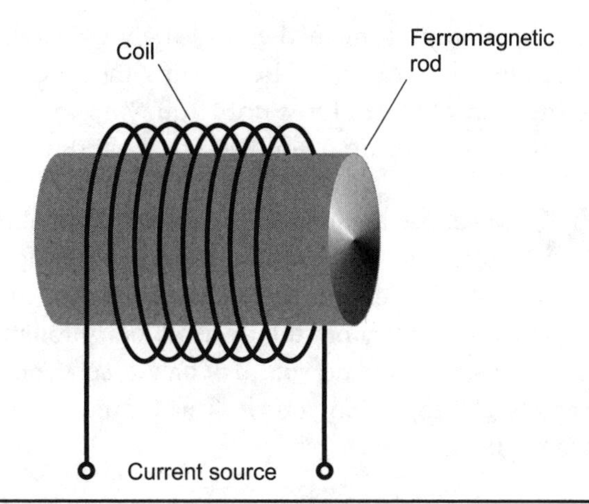

FIGURE 4-11 A simple electromagnet comprises an insulated or enameled length of wire wound around a solenoidal ferromagnetic core.

Construction

Separate the two-wire lamp cord into identical lengths of insulated wire. Place one nut at each end of a threaded steel rod. Screw the nuts in approximately 1 inch from each end of the rod. Wrap one of the wires in a neat, tight coil around the rod between the nuts, securing the wire ends to the rod outside the nuts with electrical tape, as shown in Fig. 4-12. Cut off the excess wire to make each lead 3 feet long.

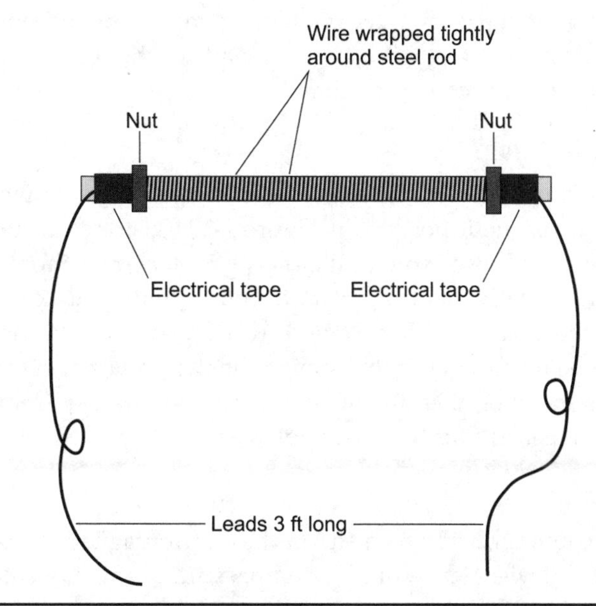

FIGURE 4-12 Construction of an electromagnet using a large threaded steel rod, two nuts, insulated wire, and electrical tape.

Strip an inch of insulation from each end of the wire. Connect one end of the coil wire to the negative battery terminal by twisting the copper strands tightly around it. Leave the other end of the coil free until you want to operate the electromagnet. You can test the device by bringing one end of the rod near various objects.

Warning! **Never use an automotive battery or other massive electrochemical battery for this experiment. The near-short-circuit produced by an electromagnet can cause the acid from such a battery to boil out and burn you. Clothing offers minimal protection; the acid will eat through fabric in a hurry. If the acid gets in your eyes, it can blind you. Use only a conventional 6-V lantern battery. Always wear safety glasses when you work with batteries and high-current devices such as electromagnets.**

Warning! **Never operate an electromagnet near any component, device, or system intended or designed for medical use.**

Warning! **Don't place an active electromagnet right up next to a computer, flash drive, removable circuit card, recording tape, or old-fashioned diskette. The electromagnet's field might corrupt the data stored on the media. (Optical discs such as CD-R suffer no harm from magnetic fields.)**

Put on your safety glasses. When you connect both ends of the coil to the battery, the rod will behave like a permanent bar magnet. Your DC electromagnet will attract objects that contain ferromagnetic material, but not objects that lack ferromagnetic material. Don't leave both ends of the coil connected to the battery for more than a few seconds at a time.

Which Pole Is Which?

You can use a trick called the *right-hand rule* to determine which end of the rod represents the magnetic north pole. Figure 4-13 illustrates the principle. If you look down the rod "endwise" so that the *conventional current* (from the positive end of the coil to its negative end) appears to rotate counterclockwise as it follows the wire, then the magnetic flux flows toward you, and the near end of the rod constitutes the north pole. If the conventional current appears to rotate clockwise as it follows the wire, then the magnetic flux flows away from you, and the near end of the rod constitutes the south pole.

Now Try This!

Once you've determined which end of the electromagnet corresponds to magnetic north, test the device near a hiker's compass. To begin, move the electromagnet at least 10 feet away from the compass. Align the compass so that its needle points toward 0° azimuth on the scale.

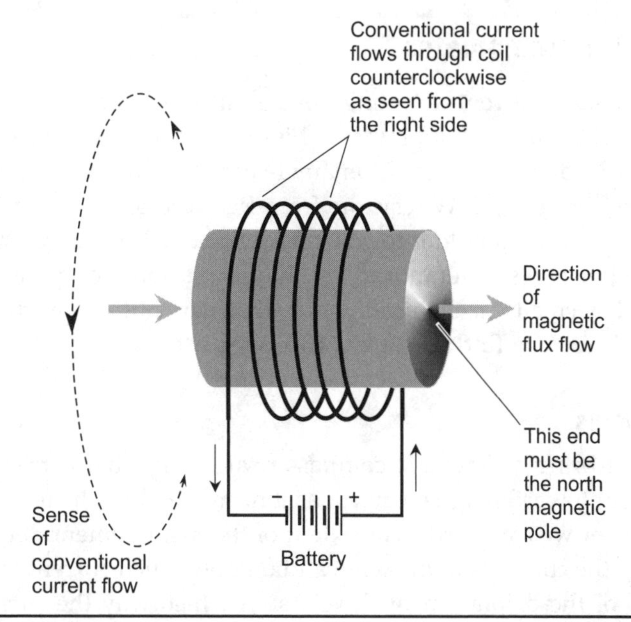

FIGURE 4-13 The right-hand rule reveals which end of a DC electromagnet behaves as the north pole.

Bring the magnetic north end of the rod near the compass from the right-hand (east) side, keeping the rod in the same plane as the compass face. Does the compass needle rotate even when the electromagnet is disconnected from the battery? If so, then you know that the rod holds some residual magnetism as a result of your earlier tests.

Connect both ends of the coil to the battery, leaving the magnetic north pole of the rod near the compass. The electromagnet's north pole will repel the north end of the compass needle and attract its south end, causing the needle to turn counterclockwise. Does this result surprise you?

Fun Factoid

The "north poles" of magnets point north because the *geomagnetic* north pole is in fact a *magnetic* south pole! Conversely, the geomagnetic south pole is actually a magnetic north pole. The geomagnetic poles attract and repel the poles of all permanent magnets and DC electromagnets. The term "north pole" arose when early experimenters floated a bar magnet on a piece of wood in a tub full of water. One end of the magnet always swung toward the north, so scientists called that end its "north pole." Even today, you can find bar magnets with the ends labeled N and S. Opposite magnetic poles attract, so if you imagine the earth as a huge magnet, the S belongs in the Arctic while the N belongs in the Antarctic!

Experiment 2: Galvanometer

You can make a current-detecting device called a *galvanometer* by wrapping some wire in a coil around a compass. For this experiment, you'll need a hiker's compass calibrated in degrees, 3 feet of enamel-coated copper wire, a sheet of fine-grain sandpaper, several 1/2-W resistors from Radio Shack or a similar outlet, six fresh AA flashlight cells, a holder for four AA cells, two holders for single AA cells, and some jumper wires to temporarily "kludge" various components together. You'll also need a small mechanical punch that can put 1/4-inch-diameter holes in cardboard. You can find one at an office supply store.

How It Works

When you bring a magnetic compass near a wire that carries DC, the compass won't point toward magnetic north as it normally does. Instead, its needle deflects to the east or west of north. The extent of the displacement depends on how close you bring the compass to the wire, and on how much current the wire carries. The direction of the displacement depends on which way the current flows through the wire.

When scientists first observed this effect, they tried different arrangements to see how much the compass needle could be displaced, and how small a current could be detected. Experimenters tried to obtain the greatest possible current-detecting sensitivity. When they wrapped the wire in a coil around the compass, they got a device that could detect (and to some extent quantify) small currents. Once the experimenters had built this apparatus, they noticed that the extent of the needle displacement increased with increasing current.

Experimenters placed a compass on a horizontal surface so that the needle pointed toward the N on the scale (*magnetic azimuth* 0°) with no current flowing in the coil. When a source of DC, such as a battery, was connected to the coil, the compass needle moved. As higher-voltage batteries were connected to increase the current in the coil, the compass needle deflection angle increased, but it never went past 90° either way. Reversing the polarity of the applied voltage reversed the sense of the needle deflection.

Construction

Wind 8-1/2 turns of enameled (not bare) copper wire around a magnetic compass so that the coil turns lie along the N-S axis of the compass as shown in Fig. 4-14. Use small-gauge, enamel-coated magnet wire of the sort available at most Radio Shack stores (part number 278-1345 at the time of this writing). To ensure that you get a mechanically stable coil, paste the compass onto a small rectangular sheet of cardboard, use a punch to put small holes in the cardboard just above the N and just below the S, and then wind the wire through the holes, passing alternately over

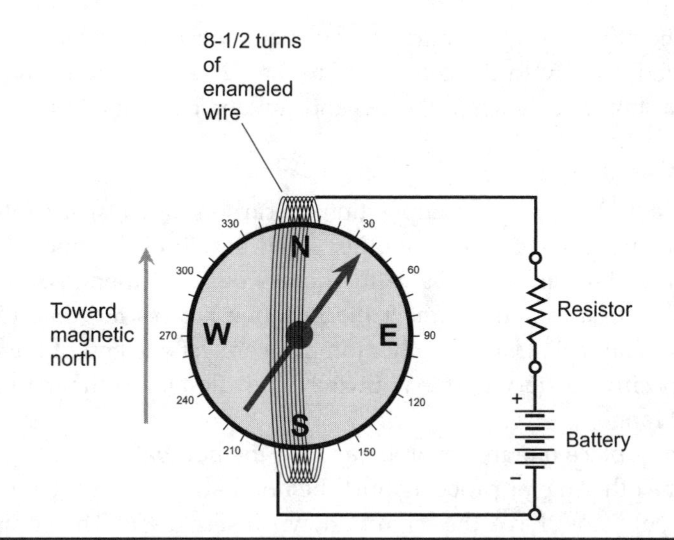

FIGURE 4-14 Galvanometer and associated circuitry. The compass must lie flat on a horizontal surface so the needle points toward N (magnetic azimuth 0 degrees) when the battery is disconnected.

and under the compass. Leave a few inches of wire protruding from each end of the coil. Figure 4-15 shows the layout that I used. Align the compass so that its needle points toward N on the scale (magnetic azimuth 0°). Make sure that the coil carries no current. If you've built the device properly, the needle should align with the coil.

Using a jumper, connect one end of the coil to the negative terminal of a single AA cell. Connect another jumper to the other end of the coil. Then, for a couple of seconds, touch the non-coil end of that jumper to the positive cell terminal. The compass needle should rotate by almost 90° so that it points either slightly north of

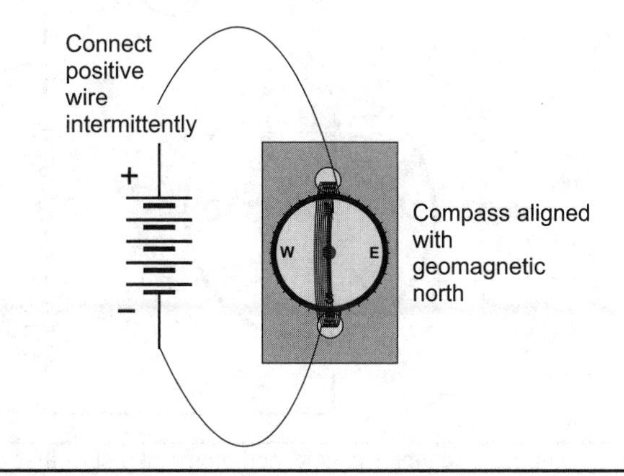

FIGURE 4-15 Galvanometer assembly. The jumper for the positive battery terminal is normally disconnected.

magnetic east or slightly north of magnetic west. Don't leave the galvanometer connected directly to the cell for more than 2 seconds at a time because the coil places a short circuit across the cell and can deplete it in a hurry!

Testing

Obtain a 1/2-W carbon-composition or carbon-film resistor rated at 680 ohms, another rated at 470 ohms, another rated at 330 ohms, and five more rated at 220 ohms. Put four AA cells in the holder built for them, so that you get a 6-V battery. With a jumper, connect the negative battery terminal to one end of the galvanometer coil. Using another jumper, connect one end of the 680-ohm resistor to the positive battery terminal. Switch your digital multimeter to a moderate DC current range.

Firmly place one meter probe against the non-battery end of the resistor, and place the other meter probe against the non-battery end of the galvanometer coil. You should now have the circuit shown in Fig. 4-16. The compass needle will probably rotate toward the east of north. If it goes west, reverse the coil connections to make the current flow the other way. If your digital meter displays negative current, reverse the probes so it shows positive current. Write down the readings from the digital meter and the compass azimuth scale.

Disconnect your digital meter and replace the 680-ohm resistor with one rated at 470 ohms. Repeat the current-versus-deflection experiment. Do the same thing with the 330-ohm resistor, and then with the 220-ohm resistor. Keep track of all your digital meter and galvanometer readings in tabular form.

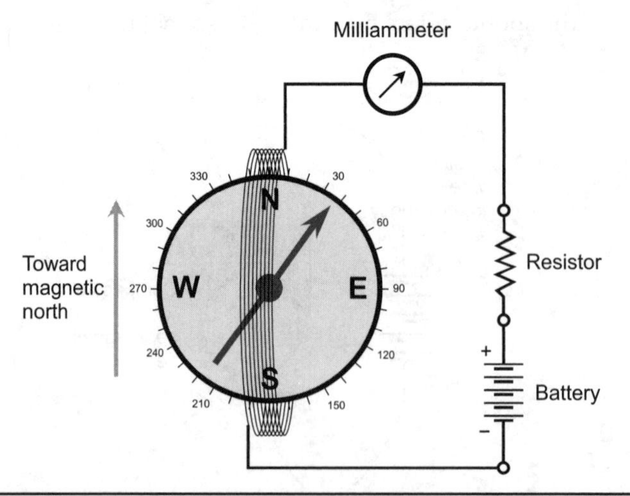

FIGURE 4-16 Arrangement for testing the galvanometer. Make sure that the compass needle points exactly toward N on the scale under no-current conditions, and deflects toward the east when current flows through the coil.

Wrap a second 220-ohm resistor in parallel with the existing one, so that you get 110 ohms. Repeat the measurements. Add a third 220-ohm resistor to the parallel combination, getting a net series resistance of approximately 73 ohms, and test the system again. Then add a fourth 220-ohm resistor in parallel, getting about 55 ohms; test again. Then add a fifth 220-ohm resistor to obtain a net resistance of 44 ohms, and test still another time.

Increase the battery voltage by taking advantage of the single-cell holders. Place an AA cell into each holder. Wire one of them up in series with the four AA cells to get a five-cell battery, and repeat the experiment with 44 ohms of resistance. Then wire the second cell in series to get a six-cell battery, and once again, do the experiment with 44 ohms of resistance.

Calibration

After you've made all the measurements and written down all the readings from your digital current meter and galvanometer, compile a table showing the number of AA cells in the first (leftmost) column, the rated resistor values in the second column, the current levels in the third column, and the compass needle deflection angles in the fourth (rightmost) column. Table 4-2 shows my results. Yours will likely differ somewhat, but they should fall "in the same ball park."

Create a calibration graph by plotting the data from your version of Table 4-2 on a coordinate grid. The horizontal axis should portray the actual current (in milliamperes), and the vertical scale should portray the compass needle deflection angle (in degrees of magnetic azimuth). Connect the dots by curve fitting to obtain

TABLE 4-2 Current Levels and Deflection Angles that I Obtained with AA Cells and Resistors in Series with a Compass-Based Galvanometer

Number of AA Cells in Series	Resistance (ohms)	Current (milliamperes)	Deflection (degrees)
0	Infinity	0	0
4	680	9.3	7
4	470	13	12
4	330	19	19
4	220	27	28
4	110	53	40
4	73	80	50
4	55	104	54
4	44	126	60
5	44	157	64
6	44	173	66

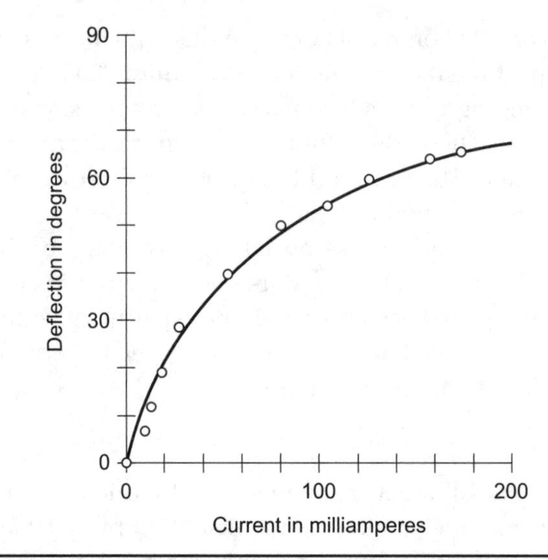

FIGURE 4-17 Compass needle deflection versus coil current. This graph reflects my experimental results, which appear in Table 4-2.

a continuous graph of deflection vs. current. Figure 4-17 shows the points and curve that I got. Yours should look similar.

Where to Find Manufactured Inductors

As of this writing, an outfit called *Alltronics* provides an excellent selection of inductors and coil cores, including pot cores. In fact, they have lots of components for the electronics hobbyist and experimenter. Visit their website at **alltronics.com**. You can also find various vendors (they morph and change as time goes by) with ads in *QST*, the magazine for radio hams published by the American Radio Relay League (ARRL). If the ARRL allows a vendor to advertise in their publications, you can have reasonable confidence that the vendor is reputable.

CHAPTER 5
Transformers

Engineers and technicians (and you!) can use *transformers* to obtain the optimum operating voltage for a circuit, device, or system from standard AC utility electricity. Transformers also have other applications. For example, they can:

- Match impedances between two different circuits so they work together in the most efficient way possible.
- Provide DC isolation between circuits, while letting AC pass, so the DC from one circuit won't affect the other circuit.
- Make balanced (double-ended) and unbalanced (single-ended) circuits, feed systems, and loads compatible with each other.

Voltage Transformation

A typical transformer contains two coils of insulated or enameled wire, along with a *core* or *form* on which they're wound. The *primary* is the coil or winding across which you input the AC. The *secondary* is the coil or winding in which AC appears as a result of the current in the primary.

Turns Ratio

Engineers define the *primary-to-secondary turns ratio* in a transformer as the number of turns in the primary T_{pri} divided by the number of turns in the secondary T_{sec}. You can denote this ratio as $T_{pri} : T_{sec}$ or T_{pri} / T_{sec}, where T stands for the number of turns.

Voltage-Transfer Ratio

In a well-built transformer, the *primary-to-secondary voltage-transfer ratio* equals the primary-to-secondary turns ratio. You can write this fact in mathematical terms as

$$E_{pri} / E_{sec} = T_{pri} / T_{sec}$$

where the E represents AC voltage.

Example 1

Suppose that a transformer has a primary-to-secondary turns ratio of 10:1. You apply 120 V AC across the primary winding. This device is a *step-down* transformer, as shown in Fig. 5-1. You can use the above equation to solve for E_{sec}. Start with

$$E_{pri}/E_{sec} = T_{pri}/T_{sec}$$

Plug in the values $E_{pri} = 120$ and $T_{pri}/T_{sec} = 10$ to obtain

$$120/E_{sec} = 10$$

Finally solve for E_{sec} to get

$$\begin{aligned} E_{sec} &= 120/10 \\ &= 12 \text{ V AC} \end{aligned}$$

Example 2

Consider a transformer with a primary-to-secondary turns ratio of 1:5. You apply 120 V AC across the primary. In this case, you have a *step-up* transformer (Fig. 5-2). Start again with

$$E_{pri}/E_{sec} = T_{pri}/T_{sec}$$

Inputting $E_{pri} = 120$ and $T_{pri}/T_{sec} = 1/5 = 0.2$, you get

$$120/E_{sec} = 0.2$$

which solves to

$$\begin{aligned} E_{sec} &= 120/0.2 \\ &= 600 \text{ V AC} \end{aligned}$$

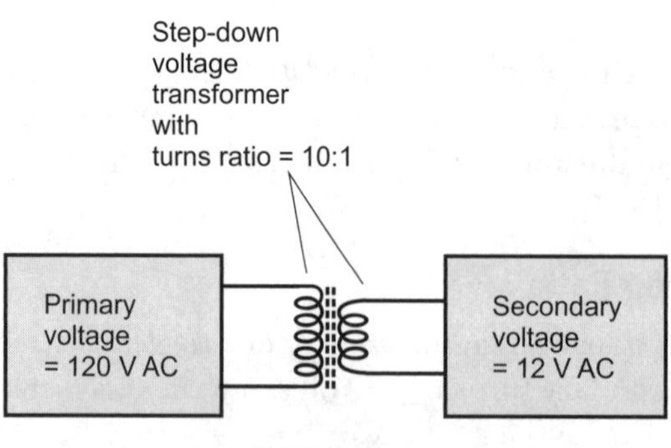

FIGURE 5-1 Illustration for Example 1.

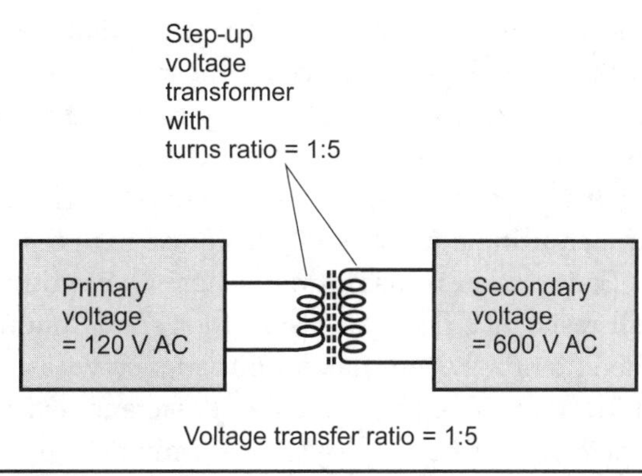

Figure 5-2 Illustration for Example 2.

Impedance Matching

In a DC circuit, the opposition to current results from resistance in the components and wiring. In an AC circuit, resistance opposes the flow of current in the same way as it does for DC. But AC circuits have another effect, *reactance*, which also opposes (or *impedes*) current. Reactance can manifest itself as capacitance or as inductance, as described in the previous two chapters. *Impedance* is a specialized mathematical combination of resistance, capacitive reactance, and inductive reactance.

> **Tip**
>
> Resistance doesn't change with AC frequency. But reactance does, if the circuit contains any inductance and/or capacitance. If you hold all other factors constant, the capacitive reactance decreases as the AC frequency rises, and the inductive reactance increases as the AC frequency rises.

An AC circuit or system works best when the impedance of a power source equals the impedance of the load into which that power goes, and neither the source nor the load contain any reactance. This ideal state of affairs does not come about by chance. Usually the two impedances must be rendered identical and reactance-free by means of transformers. This process is called *impedance matching*.

Impedance Matching Matters!

In any AC system, the load will accept all of the power only when the impedances of the load and the source are nonreactive and identical. The presence of reactance, and/or the existence of a significant discrepancy between the source and load resistances, will degrade the performance by complicating the transfer of power from the source to the load.

In a sound system, *audio transformers* can ensure that the output impedance of an amplifier is the same as the impedance of the speakers or headset, as long as both impedances are free of reactance (that is, they're *pure resistances*). The impedance is usually 4 to 16 ohms for speakers and headsets, although some communications-type headsets have impedances ranging between 600 and 2000 ohms.

Most RF transmitting equipment is designed to operate into a 50-ohm, purely resistive load. Some radio transmitting systems have output-tuning circuits that allow for small resistance fluctuations and/or small amounts of reactance in the load. But if the antenna system has an impedance whose resistive component is much more or less than 50 ohms, or if a lot of reactance exists, an *antenna tuner*, also called a *transmatch*, must be employed to eliminate the reactance and transform (match) the remaining load resistance to 50 ohms.

Matching Formula

Let Z_{load} represent the impedance of a purely resistive load that you connect to the secondary of a transformer with a primary-to-secondary turns ratio of T. You can calculate the impedance Z_{source} that appears across the transformer primary (neglecting transformer losses) as follows:

$$Z_{source} = Z_{load} T^2$$

You can also state this equation as

$$Z_{source} / Z_{load} = (T_{pri} / T_{sec})^2$$

For example, a transformer with a primary-to-secondary turns ratio of 1:2 has a primary-to-secondary impedance-transfer ratio of 1:4, and a transformer with a primary-to-secondary turns ratio of 3:1 has a primary-to-secondary impedance-transfer ratio of 9:1.

Example 3

You have an audio amplifier that you want to use with 8-ohm speakers, but the amplifier's output circuit has a nonreactive impedance of 800 ohms. You'll need a step-down audio transformer with a turns ratio of 10:1, the source being the amplifier and the load being the speaker, as shown in Fig. 5-3. (Note that the square of the 10:1 turns ratio equals 100:1 or 800:8.)

Example 4

You want to use a 200-ohm, nonreactive ham radio antenna with a transmitter designed for an output impedance of 50 ohms. You'll need a step-up RF impedance-matching transformer with a turns ratio of 1:2, the source being the transmitter and the load being the antenna, as shown in Fig. 5-4.

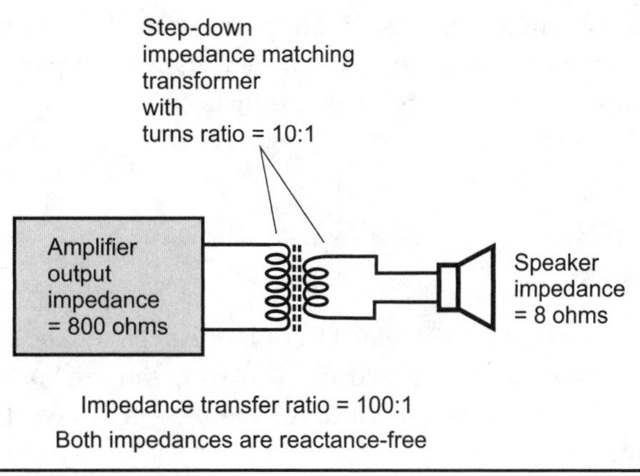

FIGURE 5-3 Illustration for Example 3.

Frequency Range

An impedance-matching transformer must be designed for a specific range of frequencies. To that end, the inductive reactances of the transformer windings should lie in the same general range as the source and load resistances at the frequencies in use.

Some impedance transformers are *wideband* in nature, meaning that they can function over a considerable span of frequencies. A *balun*, or specialized balanced-to-unbalanced transformer popular with ham radio operators, is an example of such a device.

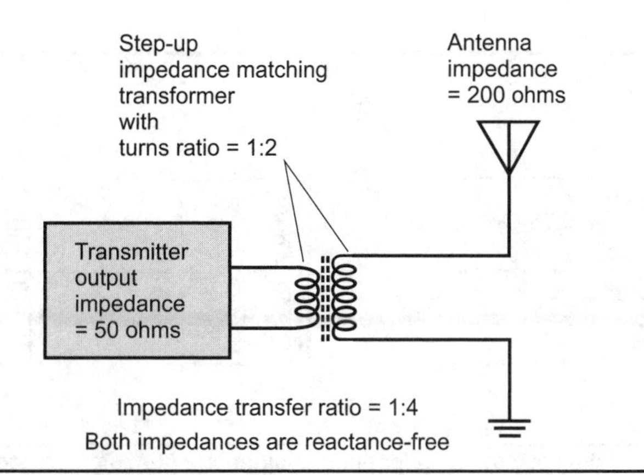

FIGURE 5-4 Illustration for Example 4.

In radio antenna systems, a good transmatch can interface loads whose impedances contain reactance, and whose resistive components vary considerably from 50 ohms. The transmatch has a transformer to match the resistances, along with an inductance-capacitance (*LC*) network that you can adjust to cancel out the load reactance.

Autotransformers

In some situations, you don't need DC isolation between the primary and secondary windings of a transformer. In a case of this sort, you can use an *autotransformer* that consists of a single, tapped winding. Figure 5-5 shows three autotransformer configurations.

1. The unit at A has an air core and operates as a step-down transformer for RF.
2. The unit at B has a laminated-iron core and operates as a step-up transformer for AC.
3. The unit at C has a powdered-iron core and operates as a step-up transformer for RF.

You'll find autotransformers in older radio receivers and transmitters, especially vintage shortwave and ham radios. Autotransformers work well in impedance-matching applications. They are occasionally, but not often, used in AF applications and in 60-Hz utility wiring.

> **Tip**
>
> In electric utility circuits, autotransformers can step the voltage down by a large factor, but they can't efficiently step voltages up by more than a few percent.

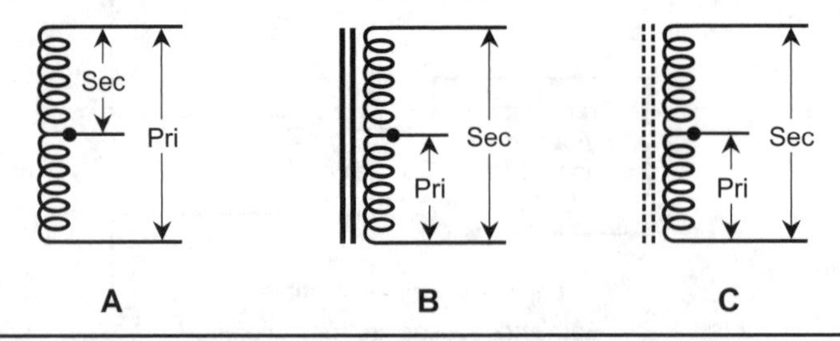

FIGURE 5-5 Schematic symbols for autotransformers. At A, air core, step-down. At B, laminated-iron core, step-up. At C, powdered-iron core, step-up.

Transformers for AC

Typical solid-state devices operate from voltages ranging from about 5 V to 50 V. For operation from 120-V AC utility mains, such equipment must have step-down transformers in their power supplies. Certain devices, such as older television (TV) sets and *vacuum-tube* systems, need step-up transformers because most vacuum tubes (called *valves* in England) require more than 120 V.

Size and Mass

Most consumer equipment comprises solid-state components exclusively, and consumes relatively little power so the transformers can be small and light. In high-powered AF or RF amplifiers, whose transistors sometimes demand more than 1000 watts (1 kW) of power, the transformers require heavy-duty secondary windings, capable of delivering current upwards of 90 A, so they're bigger and heavier.

Older TV receivers have *cathode-ray tube* (CRT) displays that need several hundred volts, derived from step-up transformers. Those transformers need not deliver a lot of current, so they have little bulk or mass, even though the output voltages are high.

Another type of system that needs high voltage is a vacuum-tube RF power amplifier, popular among ham radio operators. It might demand 2000 V (2 kV) to 5000 (5 kV) at around 500 mA from the power supply. These systems have more bulky and massive transformers.

Warning! **Treat any voltage higher than 12 V as potentially dangerous. The voltage in a TV set, or in some shortwave and ham radios, can present an electrocution hazard even after you power the system down unless the filter capacitors have effective bleeder resistors across them. Do not try to service such equipment unless you have the necessary training.**

Geometry

The properties of a transformer depend on the shape of its core and on the way in which the wires surround that core. In electricity and electronics practice, you'll encounter several different types of *transformer geometry*. The *E core*, perhaps the most common configuration, gets its name from the fact that it has the shape of a capital letter E. A bar, placed at the open end of the E, completes the core assembly (Fig. 5-6A). You can wind the primary and secondary on an E core in either of two ways.

The simpler method involves winding both the primary and the secondary around the middle bar of the E, as shown in Fig. 5-6B. Engineers call this scheme the *shell winding method*. It provides excellent *coupling* (transfer of current) between the primary and secondary, but it has significant capacitance between the primary

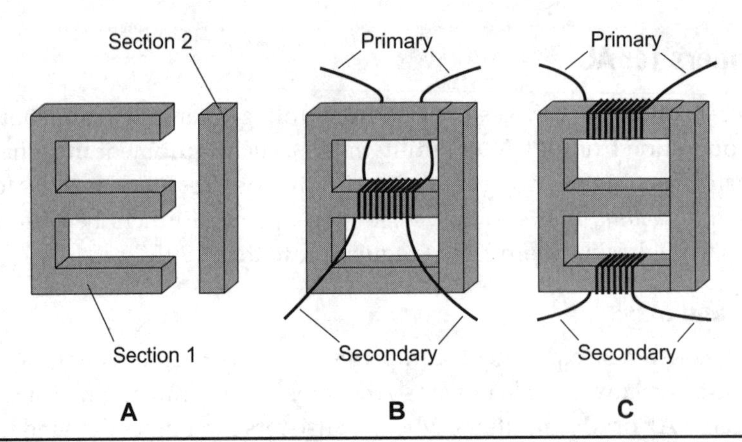

FIGURE 5-6 At A, a utility transformer E core, showing both sections. At B, the shell winding method. At C, the core winding method.

and the secondary because one winding is literally on top of the other one. (Such *interwinding capacitance* can sometimes be tolerated, but often it cannot.) Another disadvantage of the shell geometry is the fact that, when you wind coils one on top of the other, the transformer can't handle much voltage. High voltages can cause *arcing* (sparking attended by unwanted current) between the windings, which can destroy the insulation on the wires, lead to permanent short circuits, and even set the transformer on fire.

If you don't want to use the shell method, you can employ the *core method* of transformer winding. In this scheme, you place one winding at the bottom of the E section, and the other winding at the top, as shown in Fig. 5-6C. The core method has far less interwinding capacitance than a shell-wound transformer designed for the same voltage-transfer ratio because the windings are far apart. In addition, a core-wound transformer can handle higher voltages than a shell-wound transformer of the same physical size.

> **Here's a Quirk!**
> Some transformer manufacturers leave the center part of the E out of the core, producing a so-called *O core* or *D core*. These cores behave, in practice, pretty much the same as an E core does, but physically they're less robust.

Efficiency

In a real-world transformer, some power gets lost in the windings and core. *Conductor losses* occur because of the *ohmic resistance* of the wire that makes up the windings. *Core losses* occur because of *eddy currents* (circulating currents in the iron core material) and *hysteresis* (sluggishness of the core's response to AC magnetic fields). The efficiency of a transformer is, therefore, always less than 100%. The power loss appears in the form of heat; the windings and the core rise in temperature.

Let E_{pri} and I_{pri} represent the primary voltage and current in a hypothetical transformer, and let E_{sec} and I_{sec} represent the secondary voltage and current. In a perfect transformer, the product $E_{pri} I_{pri}$ would equal the product $E_{sec} I_{sec}$. But in a real-world transformer, $E_{pri} I_{pri}$ always exceeds $E_{sec} I_{sec}$. You can calculate the efficiency *Eff* of a transformer as

$$Eff = E_{sec} I_{sec} / (E_{pri} I_{pri})$$

Expressed as a percentage, the efficiency $Eff_{\%}$ is

$$Eff_{\%} = 100\ E_{sec} I_{sec} / (E_{pri} I_{pri})$$

The lost power P_{loss} in the transformer windings and core is

$$P_{loss} = E_{pri} I_{pri} - E_{sec} I_{sec}$$

Tip

Alternating current can pass through a transformer, but direct current can't. If a transformer's input contains a DC component, that component won't appear in the output. Conversely, if you impose DC across the secondary winding of a transformer, that component can't "leak back" to the primary circuit.

Did You Know?

The efficiency of a transformer varies depending on the load connected to the secondary. If the secondary circuit demands too much current, the transformer efficiency goes down. The efficiency will also suffer if a large DC component exists in the input AC because the magnetic field from the DC can partially saturate the core. Vendors rate their transformers according to the maximum amount of output power that they can deliver without serious degradation in efficiency.

Example 5

You connect a load across the secondary of an AC transformer. The voltage across the primary equals 120 V AC, and the current through the primary equals 2.57 A. You find that the voltage across the secondary is 12.3 V AC, and the current drawn by the load connected across the secondary is 19.9 A. To determine the transformer efficiency *Eff*, use the formula to get

$$Eff = E_{sec} I_{sec} / (E_{pri} I_{pri})$$
$$Eff = (12.3 \times 19.9)/(120 \times 2.57)$$
$$= 244.77/308.4$$
$$= 0.794$$

If you want to express the efficiency as a percentage $Eff_{\%}$, then you can multiply *Eff* by 100, getting $Eff_{\%} = 79.4\%$.

AF versus RF

In an AF or RF system, signals can transfer from stage to stage in various ways. Transformer coupling was once a commonly used method, although in recent years, it has fallen from favor as a result of the desire for compactness and light weight in electronic equipment. In systems that still use them, *coupling transformers* are chosen so that the output impedance of the first stage matches the input impedance of the second stage, with no reactance in either impedance. In an RF system, capacitors can be connected across the transformer windings to optimize performance at a particular frequency.

Transformer coupling still offers at least one big asset, mass obsession with compactness notwithstanding: It minimizes capacitance between adjacent stages (for example, amplifiers). This attribute is especially desirable in RF systems. In transmitter power amplifiers, the output is almost always coupled to the antenna by means of a tuned transformer. The main disadvantages of transformer coupling are the fact that it costs more than simpler coupling methods, and the fact that it contributes to a heavier, bulkier piece of equipment.

Audio transformers resemble those employed for 60-Hz electricity, except that the operating frequency is higher, and audio signals exist over a range, or *band*, of frequencies rather than at a single frequency. Most AF transformers are constructed like miniature utility transformers. They have E cores with primary and secondary windings wound around the crossbars. Audio transformers can function in either the step-up or step-down mode, and are designed to match impedances rather than to produce specific output voltages.

Most RF transformers have primary and secondary windings, just like utility transformers. In construction, you can use powdered-iron cores up to quite high frequencies. Toroidal cores work especially well because of their *self-shielding* characteristic (all of the magnetic flux stays within the core material). The optimum number of turns depends on the frequency, and also on the permeability of the core.

In high-power RF applications, many engineers prefer to use air-core coils in transformers. Although air has low permeability, it has negligible loss, and will not heat up or fracture as powdered-iron cores sometimes do. However, some of the magnetic flux extends outside of an air-core coil, potentially degrading the performance of the transformer when it must function in close proximity to other components.

Tip

A major advantage of coil-type RF transformers, especially when wound on toroidal cores, lies in the fact that you can get them to function efficiently over a wide band of frequencies, such as from 3.5 MHz to 30 MHz. A transformer designed to work well over a sizable frequency range is called a *broadband transformer*.

Experiment 1: Transformer Tests

For this and the following experiment, you'll need two Enercell power adapters, available at most Radio Shack retail outlets as part number 273-0331. These adapters convert 120 V AC to either 24 V AC or 18 V AC (*not DC!*)—switch-selectable, according to the manufacturer. You'll also need your digital multimeter, along with a short, two-wire utility extension cord with a three-outlet block tap that you'll chop off! Figure 5-7 shows these components. In addition, you'll need a diagonal cutter that can double as a wire stripper, and some electrical tape.

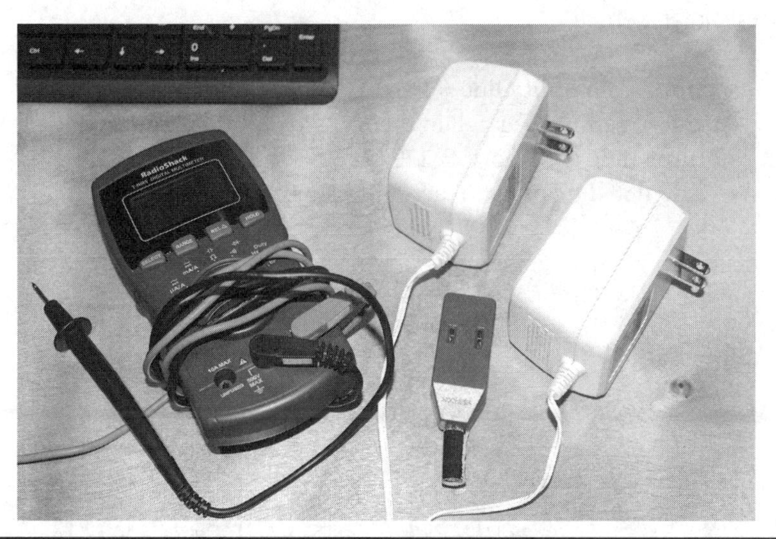

Figure 5-7 Devices for transformer experiments: Multimeter, hacked extension cord outlet, and two adapters that convert 120-V AC utility electricity to either 18 V AC or 24 V AC (Radio Shack part number 273-0331 or equivalent).

Measure the Voltages

The Enercell adapters have small connectors on their output cords. Remove these connectors with your diagonal cutter. Then separate the cut-off wires for a few inches "upstream" and strip 1 inch (2.5 centimeters) of insulation from the ends of both wires.

Plug the adapters into standard utility outlets, making certain that the exposed wire ends don't short out or touch any other conducting object. Using your digital multimeter, measure the actual input and output voltages at the "18 V" and "24 V" settings. Twist the wire ends around the meter probes and secure the connections with electrical tape. (Polarity doesn't matter here because you're working with AC.) Alternatively, you can "stuff" the adapter output wire ends into the holes where the probes connect to the meter.

Now measure the actual utility input voltage at the wall outlet. Make sure your multimeter is set to measure AC volts, not DC volts! Once you know the input and output voltages, you can calculate the exact primary-to-secondary turns ratios, which equal the input voltage divided by the output voltage.

My Results

When I conducted these tests, I got identical results for both units. The actual output voltages exceeded the quoted values, probably because I had not connected any output loads to the transformers. (Evidently, the manufacturer rates the output voltages for these modules on the assumption that users will connect them to moderate loads.) Here are the voltages I observed under no-load conditions.

High-Voltage Setting

- Input from utility line = 118.0 V AC
- "24 V" setting (Fig. 5-8 in theory) yields actual output of 27.9 V AC
- Turns ratio (in theory) = 120/24 = 5 : 1
- Calculated turns ratio (actual) = 118.0/27.9 = 4.23 : 1

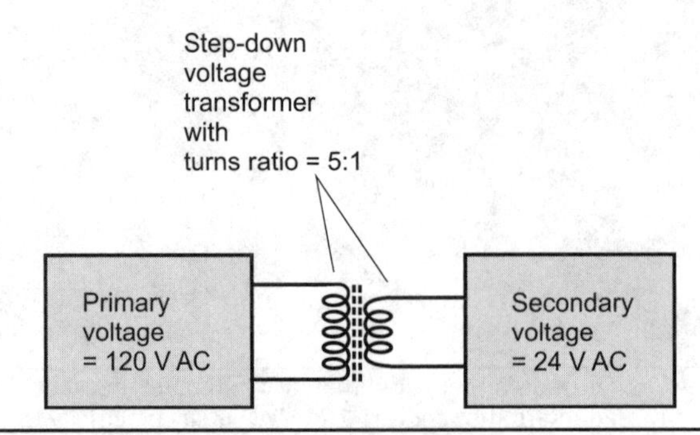

FIGURE 5-8 Step-down transformer for 120 V AC to 24 V AC.

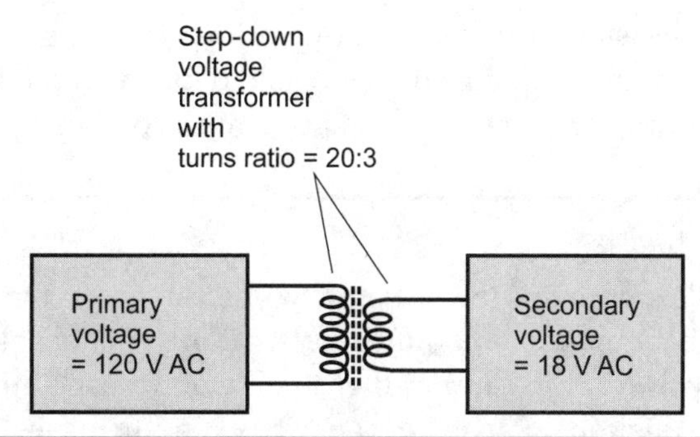

FIGURE 5-9 Step-down transformer for 120 V AC to 18 V AC.

Low-Voltage Setting

- Input from utility line = 118.0 V AC
- "18 V" setting (Fig. 5-9 in theory) yields actual output of 20.4 V AC
- Turns ratio (in theory) = 120/18 = 6.67 : 1
- Calculated turns ratio (actual) = 118.0/20.4 = 5.78 : 1

Now Try This!

Connect a load to the transformer output so that the device must deliver some current. I suggest a conventional incandescent utility bulb, say 15 W. Of course, the bulb, designed for use at 120 V AC, won't glow at full brilliance, but it will create a significant load for the transformer. To connect the bulb to the transformer output, use a cheap desk lamp. Cut off the lamp's plug and strip about 1 inch (2.5 centimeters) of insulation from the ends of the cord wires. Twist those wires around the output wires of one of your transformers.

What do you think will happen to the transformer output voltage under load? Measure it and see! Try bulbs that demand more current, such as 25 W, 40 W, or 60 W. Does the output voltage drop as you should expect? If so, does it drop more and more as the load demands more current?

Experiment 2: Back-to-Back Transformers

The Enercell engineers designed their power adapter module to work as a step-down transformer. But as a "tweak freak," I wondered whether or not I could connect two of those modules back-to-back (the secondary of one going to the secondary of the other) and get high voltage at the primary of the other. I knew better than to plug a module's secondary into a wall outlet; such a stupid action would probably burn the winding out and might even set the module on fire! But what would happen if I applied 18 V AC or 24 V AC to the secondary of the second module and measured the voltage at its primary? I expected to find high voltage there, and I expected that the voltage would depend on the relative switch settings of the two modules. As things worked out, my hypothesis proved correct.

Test Your Theories!

Until I did experiments to verify my hypothesis about the output voltage from a back-to-back pair of modules, my notion remained theoretical. If, as an engineer, you want to *know* (rather than merely suspect) what will happen in a certain situation, you must conduct experiments or observations in the real world. As my physics advisor in college said back in the 1970s, "One experimentalist can keep a dozen theorists busy." That statement remains true today. All the theories in the Cosmos mean nothing without experimental backup. That's why labs and workshops exist!

Severed Block Tap

If you connect two transformers back-to-back, you'll get dangerous high-voltage AC on the exposed prongs of the second module unless you devise some means to convert the male plug into a female outlet. You can ensure your safety by cutting off the three-outlet block tap at the end of a cheap utility extension cord, inserting the second transformer's utility plug into the side of the block with two outlets, and then using the other side of the block as a safe spot into which you can insert your meter probe tips. Figure 5-10 shows how I did this trick. Make sure that the

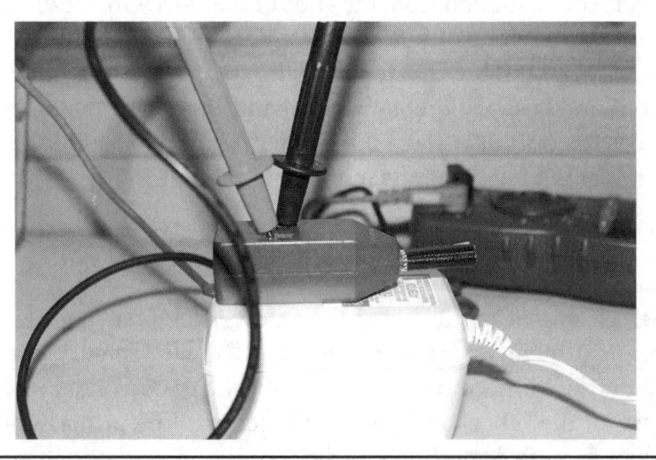

Figure 5-10 A severed block tap keeps high voltage from exposed prongs. You can insert the meter probe tips into the outlet slots.

chopped-off cord wires don't short out, and then wrap the end of the cut-off cord with electrical tape.

My Results

When I conducted these tests, the output voltage depended on the relative switch settings. With both switches set to 18 V AC or 24 V AC, I saw an output voltage slightly less than the utility voltage, doubtless the result of transformer losses. No surprise there! When the switch settings differed, I got an output voltage considerably higher or lower than the utility voltage. No surprise there either! Here are the results in full.

Switch Settings High-High

- Input from utility line = 118.0 V AC
- First unit set at "24 V" (Fig. 5-8) and second unit set at "24 V" (Fig. 5-11)
- Output = 116.7 V AC

Switch Settings High-Low

- Input from utility line = 118.0 V AC
- First unit set at "24 V" (Fig. 5-8) and second unit set at "18 V" (Fig. 5-12)
- Output = 156.3 V AC

Switch Settings Low-High

- Input from utility line = 118.0 V AC
- First unit set at "18V" (Fig. 5-9) and second unit set at "24 V" (Fig. 5-11)
- Output = 85.7 V AC

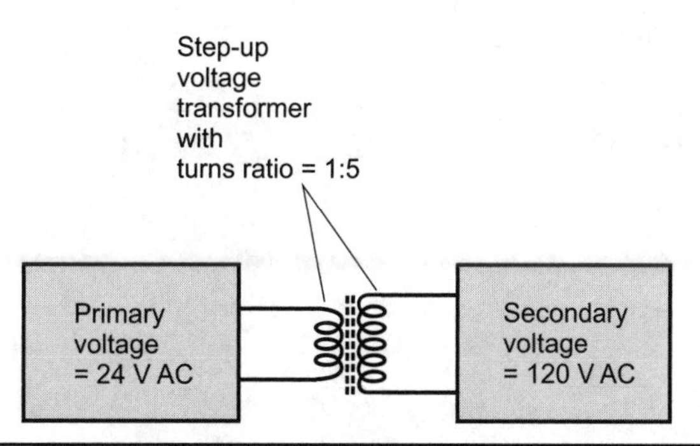

Figure 5-11 Step-up transformer for 24 V AC to 120 V AC.

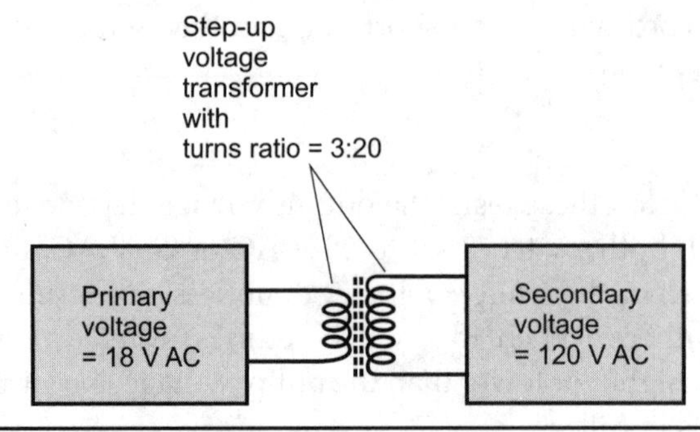

FIGURE 5-12 Step-up transformer for 18 V AC to 120 V AC.

Switch Settings Low-Low

- Input from utility line = 118.0 V AC
- First unit set at "18 V" (Fig. 5-9) and second unit set at "18 V" (Fig. 5-12)
- Output = 116.7 V AC

Warning! **Don't connect any load to the output of the back-to-back transformer combination. You might be tempted to try it, for example to plug a desk lamp into the block tap on the second transformer, but I recommend that you resist that urge. One or both of the modules could overheat or short out if the joined secondary windings are forced to carry very much current. You'll face a fire risk in that case!**

CHAPTER 6
Diodes

The term *diode* means "two elements." In the olden days of electronics (before about 1960), two-element vacuum tubes served as diodes. The *cathode* element emitted electrons, and the *anode* element picked them up. Electrons flowed easily through the tube from the cathode to the anode, but not the other way. Nowadays, most diodes comprise silicon or other semiconducting materials, but engineers still call the corresponding elements the cathode and the anode.

Characteristics

Two categories of semiconductors prevail in electronics: *N type*, in which the charge carriers are mainly electrons; and *P type*, in which the charge carriers are primarily *holes* (atoms with "missing" electrons).

The Junction

When samples of N type and P type semiconductor material rest in physical contact, you get a *P-N junction* that forms a *semiconductor diode*. Figure 6-1 shows the schematic symbol for this component. The short, straight line represents the N type material, which acts as the cathode. The arrow represents the P type material, which acts as the anode.

In a diode, electrons can normally flow against the arrow, but not in the direction that the arrow points. If you interconnect a battery, a resistor, and a diode, as shown in Fig. 6-2A, current will flow in the circuit (except when the voltage is below the *forward breakover* point, discussed in the next paragraph). Then the diode operates in a state of *forward bias* where the cathode is negative and the anode is positive.

If you reverse the battery polarity, as shown in Fig. 6-2B, no current will flow (except when *avalanche breakdown* occurs, also described in the next paragraph) because a *depletion region* forms on either side of the P-N junction where almost no electrons or holes exist. Because it lacks charge carriers that can conduct current, that zone acts as an electrical insulator or dielectric. Then the diode operates in *reverse bias* in which the cathode is positive and the anode is negative.

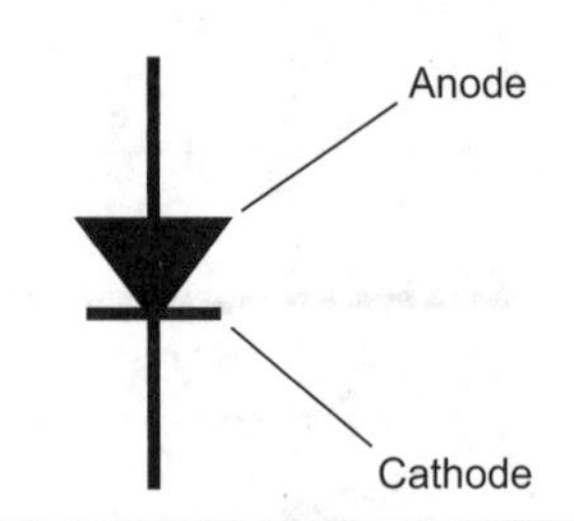

Figure 6-1 The schematic symbol for a semiconductor diode.

Breakover and Avalanche

You must apply a certain minimum forward bias, called the *forward breakover voltage*, to a diode to make its P-N junction conduct current. Depending on the type of semiconductor material, the forward breakover voltage varies from about 0.3 V to 1 V. A silicon diode has a forward breakover of 0.5 V to 0.6 V under most operating conditions.

If you reverse-bias a diode at a gradually increasing voltage, the P-N junction will conduct at and above a certain voltage called the *avalanche voltage*. That's called the *avalanche effect*. The avalanche voltage varies among different diodes. It's nearly always far greater than the forward breakover voltage.

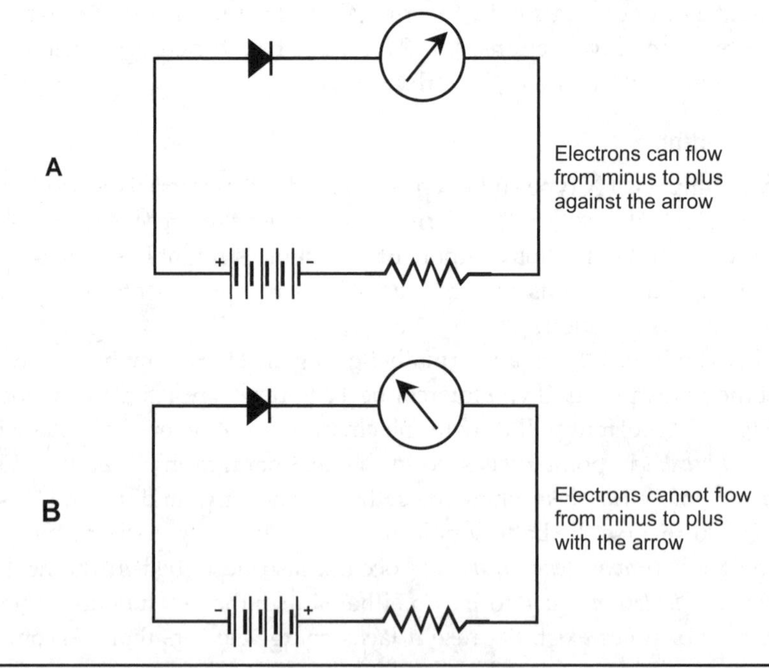

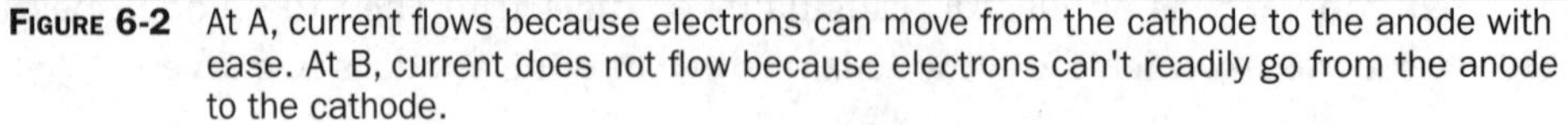

Figure 6-2 At A, current flows because electrons can move from the cathode to the anode with ease. At B, current does not flow because electrons can't readily go from the anode to the cathode.

In rectifier diodes intended for power supplies, you don't want the avalanche effect to occur during any part of the cycle, not even for the briefest instant. You can prevent that situation by choosing diodes with an avalanche voltage of at least twice, and preferably three or four times, the *peak inverse voltage* (PIV), or the highest instantaneous voltage of the AC waveform.

> **Did You Know?**
>
> A component known as a *Zener diode* actually makes good use of the avalanche effect. Zener diodes are manufactured to have precise avalanche voltages. They form the basis for *voltage regulation* in some power supplies because they limit the voltage that can exist across them under conditions of reverse bias.

Junction Capacitance

When reverse-biased, a P-N junction behaves as a capacitor because the depletion region acts as a dielectric. You can vary the *junction capacitance* by changing the reverse-bias voltage because this voltage affects the width of the depletion region. As the reverse voltage increases, the depletion region grows wider, so the capacitance gets smaller. A *varactor diode* is made with this property in mind. Typical capacitance values are on the order of a few picofarads.

Rectification

The "one-way current gate" property makes certain diodes useful for changing AC to DC. You can find *rectifier diodes* in various sizes, some for high voltage and others for low voltage; some can handle high current and others, not so much! Most rectifier diodes are silicon based. A few are made from selenium. Important features of a rectifier diode include the *average forward current* (I_o) rating and the *peak inverse voltage* (PIV) rating.

Average Forward Current

If the current through a diode gets too high, the resulting heat will destroy the P-N junction. When designing a power supply, you should use diodes with an I_o rating of at least 1.5 times the expected average DC forward current. If this current is 4 A, for example, the rectifier diodes should be rated at $I_o = 6$ A or more. You might even use 8-A diodes to remain on the safe side! But it would be silly to use a 100-A diode in a circuit in which the average forward current is 4 A. (Over-engineering is good up to a point, beyond which it enters the realm of the ridiculous!)

In a rectifier circuit, the I_o specification refers to the current that flows *within the diodes*, not the current that flows in other parts of the circuit. The current drawn by the load of a power supply often differs from the internal diode current. Also, note that I_o represents an *average* figure. The *peak* forward current can rise to 15 or

20 times the average forward current, depending on the nature of the power-supply filter that gets rid of the pulsations in rectified DC.

You can connect two or more identical diodes in parallel to increase the I_o rating of the set over that of a single component. In that case, you should install a small-value resistor in series with each diode to prevent *current hogging*, in which one diode might carry most of the current, suffer the greatest burden, and burn out as a result. You should choose the ohmic values so that about 1 V exists across each resistor.

Peak Inverse Voltage

The PIV rating of a diode is the highest instantaneous inverse (or reverse) voltage that it can withstand without avalanche effect. A good power supply has diodes whose PIV ratings significantly exceed the peak voltage of the AC that they're rectifying. If the PIV rating is too low, a diode will conduct for part of the reverse cycle, degrading the efficiency of the power supply because reverse current, even if it flows for only a tiny fraction of the cycle, "bucks" forward current.

You can connect multiple identical diodes in series to get a higher PIV capacity than a single diode alone can provide. This scheme is sometimes used in high-voltage supplies, such as those needed for tube-type amateur radio linear amplifiers. You can place a resistor, rated at about 500 ohms for each peak-inverse volt, across each diode to distribute the reverse bias voltage equally among the diodes.

Detection

One of the earliest diodes, known as a *cat's whisker*, comprised a thin wire in contact with a fragment of the mineral *galena*. This contraption could rectify weak RF currents. When experimenters connected the cat's whisker in a configuration, such as the circuit of Fig. 6-3, the resulting device picked up amplitude-modulated (AM) radio signals and produced audible output in the headset. The diode recovered the

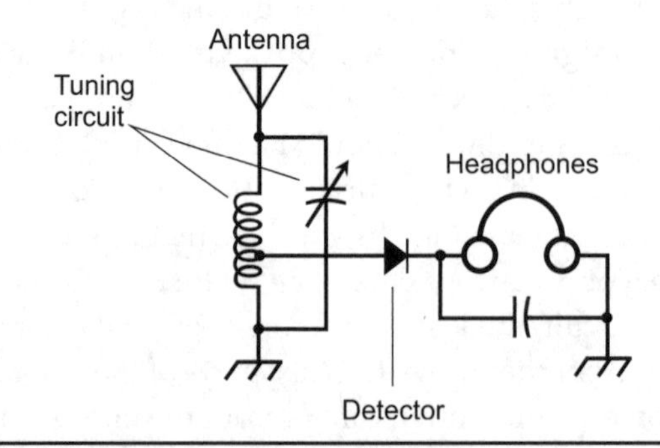

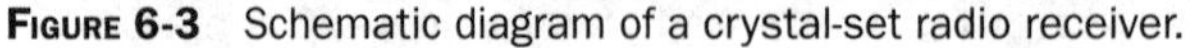

Figure 6-3 Schematic diagram of a crystal-set radio receiver.

audio from the radio signal. Engineers decided to call this process *detection* or *demodulation*.

The galena fragment, also called a *crystal*, gave rise to the moniker *crystal set* for this receiver. You can build one using an RF diode, a coil, a tuning capacitor, a headset, and a wire antenna. The circuit needs no electricity whatsoever! If a broadcast station exists within a few miles of the antenna, the demodulated RF signal alone can produce enough audio to drive the headset. For ideal performance, you can shunt the headset with a capacitor of large enough value to short out RF current to ground, but not so large as to short out the audio signal that you want to hear.

Tip

If you want a diode detector to work, you must use a diode with junction capacitance low enough so that it can rectify at the high signal frequency (often in the megahertz range) without acting like a capacitor at that frequency.

Frequency Multiplication

When AC passes through a diode, half of the cycle gets cut off. This effect occurs regardless of the applied-signal frequency as long as the diode capacitance remains small, and as long as the reverse voltage remains below the avalanche threshold. The output wave from the diode looks much different than the input wave. Engineers call this condition *nonlinearity*. Whenever a circuit exhibits nonlinearity, *harmonics* appear in the output. Harmonics are signals at whole-number multiples of the input frequency, known as the *fundamental frequency*.

To prevent harmonics or distortion from arising in a circuit, you'll want to make it *linear* so that the output waveform has the same shape as the input waveform (even if the signals aren't equally strong). But you'll sometimes *want* a circuit to act in a nonlinear fashion, such as when you intend to generate harmonics. You can use diodes to deliberately introduce nonlinearity into a circuit to obtain *frequency multiplication*. Figure 6-4 illustrates a simple *frequency multiplier* that uses an RF

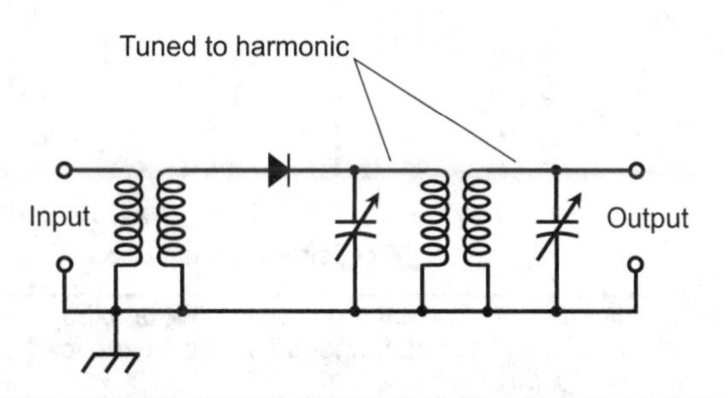

FIGURE 6-4 A frequency-multiplier circuit that uses an RF diode.

diode to generate harmonics. You should tune the output *LC* circuit to the desired harmonic frequency.

Signal Mixing

When you combine or *mix* two waves of different frequencies in a nonlinear circuit, you get new waves at frequencies equal to the sum and difference of the frequencies of the inputs. One or more diodes can provide the nonlinearity that you need to make this happen.

Imagine two RF signals with frequencies f_1 and f_2. Assign f_2 to the wave with the higher frequency, and f_1 to the wave with the lower frequency. If you combine these two signals in a nonlinear circuit called a *mixer*, you get two new signals. One of them has a frequency of $f_2 + f_1$, and the other one has a frequency of $f_2 - f_1$. Engineers call these new frequencies the *beat frequencies* and the signals *mixing products* or *heterodynes*.

Figure 6-5 shows hypothetical input and output signals for a mixer on a *frequency-domain* display. The amplitude (on the vertical scale) is a function of the frequency (on the horizontal scale). You'll see this sort of display when you look at the screen of a lab instrument, known as a *spectrum analyzer*. In contrast, an ordinary oscilloscope displays amplitude (on the vertical scale or axis) as a function of time (on the horizontal scale or axis), so it shows you a *time-domain* display.

> **Tip**
>
> For a diode to function as a frequency multiplier or mixer for RF signals, it must be of the sort that will also work as a detector at the same frequency. It should act like a rectifier, but not like a capacitor.

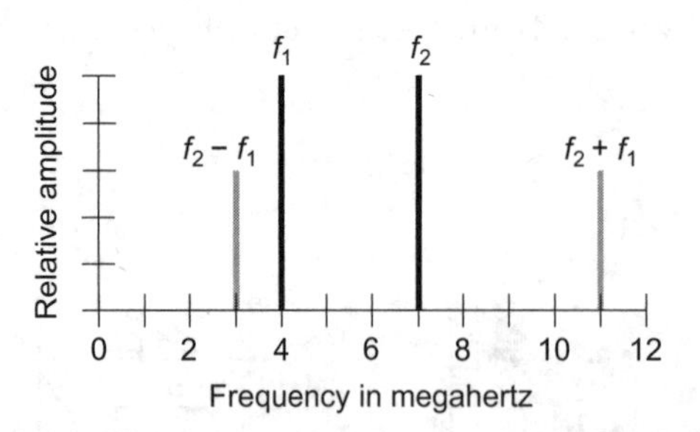

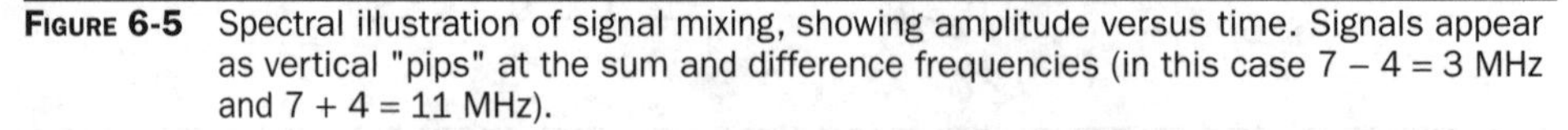

Figure 6-5 Spectral illustration of signal mixing, showing amplitude versus time. Signals appear as vertical "pips" at the sum and difference frequencies (in this case 7 – 4 = 3 MHz and 7 + 4 = 11 MHz).

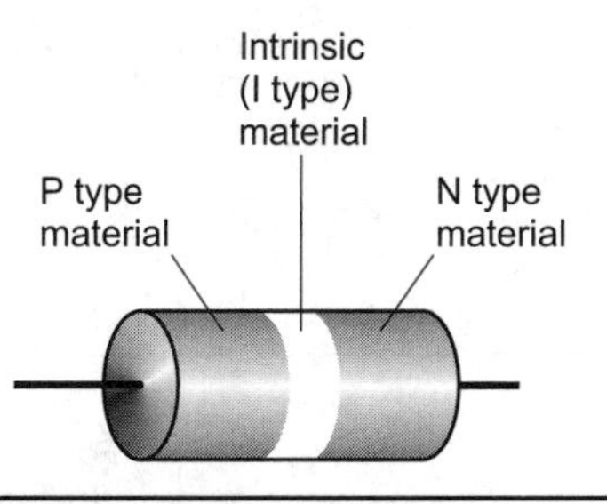

Figure 6-6 A PIN diode has a layer of intrinsic (I type) semiconductor material between the P and N type layers, decreasing the junction capacitance, as compared with a conventional RF diode.

Switching

The ability of diodes to conduct current when forward-biased and block current when reverse-biased makes them useful for switching in some applications. Diodes can perform on/off switching at millions of operations per second, compared with mechanical relays that can work at only a few hundred operations per second. Diodes offer better reliability than relays, they last longer, they take up less space, they weigh less, and they cost less.

One type of diode, made for use as an RF switch, has a special semiconductor layer sandwiched in between the P type and N type material. The material in this layer is called an *intrinsic* (or *I type*) *semiconductor*. The *intrinsic layer* (or *I layer*) reduces the capacitance of the diode by acting as a depletion zone, allowing the device to function effectively at higher frequencies than an ordinary diode can do. A diode with an I type semiconductor layer sandwiched in between the P and N type layers is called a *PIN diode* (Fig. 6-6).

> **Tip**
>
> Variable DC bias, applied to one or more PIN diodes, lets you guide or "channel" RF current to desired points without mechanical relays. A PIN diode also makes a good RF detector at frequencies of 30 MHz and above.

Voltage Regulation

Most diodes have an avalanche voltage that's higher than the reverse-bias voltage ever gets. But not all! The exact value depends on the construction of the diode, and on the characteristics of the semiconductor materials that compose it. *Zener*

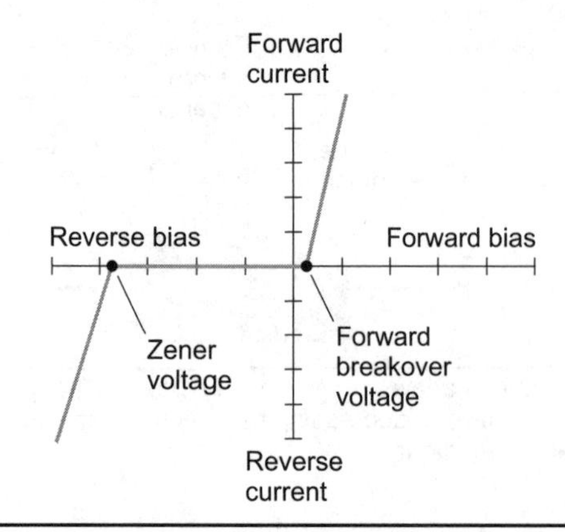

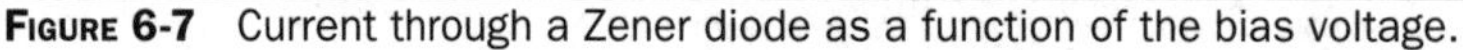

FIGURE 6-7 Current through a Zener diode as a function of the bias voltage.

diodes are specifically manufactured to exhibit well-defined, constant avalanche voltages.

Imagine that a Zener diode has an avalanche voltage, also called the *Zener voltage*, of 50 V. If you apply reverse bias to the P-N junction, the diode acts as an open circuit as long as the voltage between the P and N type materials remains less than 50 V. But if the reverse-bias voltage reaches 50 V, even for a moment, the diode conducts. This phenomenon prevents the instantaneous reverse-bias voltage from exceeding 50 V.

Figure 6-7 shows a graph of the current through a hypothetical Zener diode as a function of the voltage. The Zener voltage shows up as an abrupt rise in the current as the reverse bias increases (moving toward the left along the horizontal axis).

Figure 6-8 shows a simple Zener-diode *voltage-regulator* circuit. You should connect the cathode to the positive pole and the anode to the negative pole, opposite from the way you would hook up a diode to work as a rectifier. The series-connected

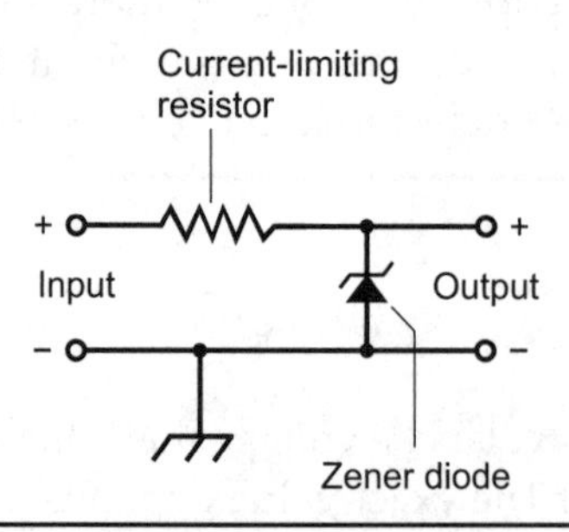

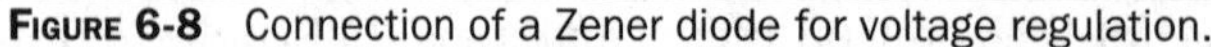

FIGURE 6-8 Connection of a Zener diode for voltage regulation.

resistor limits the reverse current that can flow through the diode. Without that resistor, the diode might conduct too much current and burn out.

Amplitude Limiting

A forward-biased diode won't conduct until the voltage reaches the forward breakover point, but it will always conduct when the forward-bias voltage exceeds forward breakover. In a diode, the voltage between the P and N type layers remains roughly equal to the forward breakover voltage as long as current flows. For silicon diodes, this *voltage drop* is 0.5 V to 0.6 V. For germanium diodes, the voltage drop is roughly 0.3 V, and for selenium diodes it's around 1 V.

You can take advantage of the "constant-voltage-drop" property of diodes to build a circuit that limits, or *clips*, the amplitude of a signal. Figure 6-9A shows how to connect two identical diodes back-to-back in parallel with the signal path to clip the positive and negative peaks. The positive and negative peak voltages can never exceed the forward breakover voltage of the diodes.

Figure 6-9B shows the input and output waveforms of a clipped AC signal. You'll sometimes find this type of circuit in the audio output circuits of older Morse code radio receivers that lack automatic gain control (AGC). The clipper prevents strong signals from "blasting" your ears with the volume turned up.

> **Tip**
>
> The downside of a *diode voltage-limiter*, such as the one shown in Fig. 6-9A, is the fact that it introduces distortion when clipping occurs. With AM signals having peaks that rise past the limiting voltage, clipping distortion can make voices difficult to understand, and it ruins the sound quality of music! However, a clipper like this one can work fine for Morse code or frequency-shift-keyed signals, where fidelity doesn't matter.

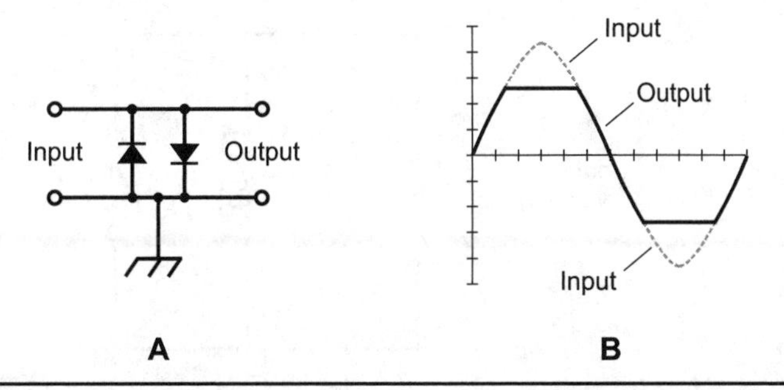

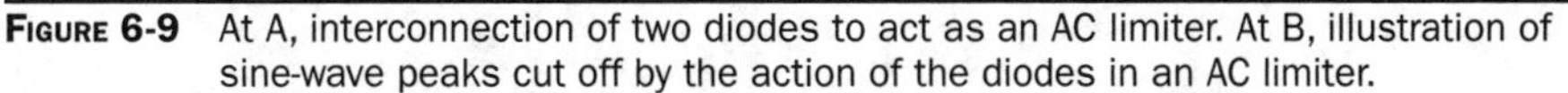

Figure 6-9 At A, interconnection of two diodes to act as an AC limiter. At B, illustration of sine-wave peaks cut off by the action of the diodes in an AC limiter.

Frequency Control

When you reverse-bias a diode, you create a depletion region at the P-N junction with dielectric (insulating) properties and the near-absence of charge carriers. The width of the depletion region depends on the reverse-bias voltage.

As long as the reverse bias remains lower than the avalanche voltage, varying the bias affects the width of the depletion region. The fluctuating width, in turn, varies the junction capacitance. This capacitance, normally a few picofarads, varies inversely with the *square root* of the reverse-bias voltage—again, as long as the reverse bias remains less than the avalanche voltage. For example, if you quadruple the reverse-bias voltage, the junction capacitance drops to half; if you decrease the reverse-bias voltage by a factor of 9, then the junction capacitance increases by a factor of 3.

Some diodes are manufactured specifically to function as variable capacitors. Such devices are known as *varactor diodes*. Varactors find their niche in a special type of circuit called a *voltage-controlled oscillator* (VCO). Figure 6-10 shows an example of a parallel-tuned *LC* circuit in a VCO using a coil, a fixed capacitor, and a varactor. The fixed capacitor (whose value should greatly exceed the capacitance of the varactor) keeps the coil from short-circuiting the control voltage across the varactor.

> **Tip**
> The schematic symbol for a varactor diode has two lines on the cathode side, as opposed to one line on the symbol for a conventional diode. You can identify it that way!

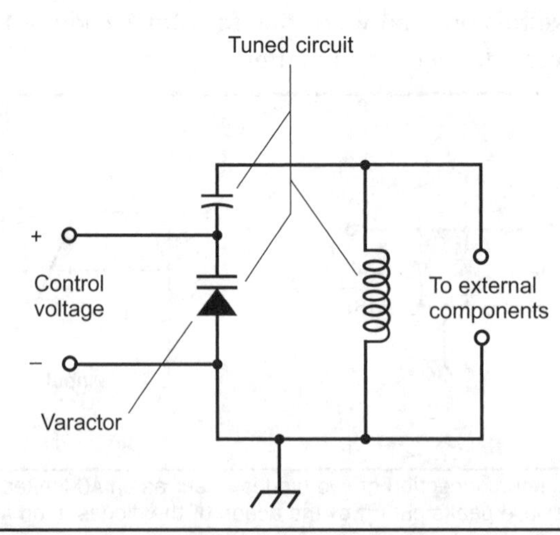

Figure 6-10 Connection of a varactor diode in a tuned circuit.

Oscillation and Amplification

Under certain conditions, diodes can generate or amplify *microwave* RF signals—that is, signals at frequencies in the multigigahertz range. Devices designed for these purposes include the *Gunn diode*, the *IMPATT diode*, and the *tunnel diode*.

Gunn Diodes

A Gunn diode can produce 100 mW to 1 W of RF power output, and is manufactured with the semiconductor compound *gallium arsenide* (GaAs). A Gunn diode oscillates because of the *Gunn effect*, named after the engineer *J. Gunn* who first observed the phenomenon in the 1960s while working for International Business Machines (IBM) Corporation.

In a Gunn diode, oscillation takes place as a result of a quirk called *negative resistance*, in which an increase in the applied voltage causes a decrease in the current flow under specific conditions. It's not true negative resistance, of course, (the closest thing to that would come from a battery!) but rather a form of current-versus-voltage behavior that runs contrary to the norm.

Here's a Twist!

A Gunn-diode oscillator, connected directly to a horn-shaped antenna, gives you a device known as a *Gunnplexer*. Amateur-radio experimenters use Gunnplexers for low-power wireless communication at frequencies of 10 GHz and above.

IMPATT Diodes

The acronym *IMPATT* comes from the first letters of the words in an arcane technical expression: *imp*act *a*valanche *t*ransit *t*ime. This type of diode, like the Gunn diode, works because of the negative resistance phenomenon. An *IMPATT diode* oscillates at microwave frequencies, just as a Gunn diode does, except that it's manufactured from silicon rather than gallium arsenide. An IMPATT diode can also operate as an amplifier for a microwave transmitter that employs a Gunn-diode oscillator. As an oscillator, an IMPATT diode produces about the same amount of usable RF output power, at comparable frequencies, as a Gunn diode does.

Tunnel Diodes

Another type of diode that can oscillate at microwave frequencies is the *tunnel diode*, also known as the *Esaki diode*. Made from GaAs semiconductor material, the tunnel diode produces minimal RF power, only enough for use in a local oscillator for a receiver or transceiver. Tunnel diodes can also work well as weak-signal amplifiers in microwave receivers because they generate very little RF noise.

Energy Conversion

Certain diodes emit radiant energy when current passes through the P-N junction in a forward direction. This phenomenon occurs as electrons "fall" from higher to lower energy states within the atoms. Other diodes can change characteristics depending on how much radiant energy strikes them. A few are designed to produce electricity directly from radiant energy.

LEDs and IREDs

Depending on the mixture of semiconductors used in manufacture, visible light of almost any color can be produced by forward-biased diodes. Infrared (IR) devices also exist; an *infrared-emitting diode* (IRED) produces energy at wavelengths slightly longer than those of visible red light, but shorter than microwave radio signals.

The intensity of the radiant energy from an LED or IRED depends on the forward current. As the current rises, the brightness increases, but only up to a certain point. If the current continues to rise, no further increase in brilliance takes place, and engineers say that the LED or IRED is working in a state of *saturation*.

Silicon Photodiodes

A silicon diode, housed in a transparent case and constructed so that IR or visible light can strike the barrier between the P and N type materials, forms a *silicon photodiode*. If you apply a reverse-bias voltage to the device at a certain level below the avalanche threshold, no current flows when the junction remains in darkness, but conduction will occur if sufficient radiant energy strikes it.

At constant reverse-bias voltage, the current varies in direct proportion to the intensity of the incoming radiant energy, within certain limits. When IR or visible light of variable intensity strikes the P-N junction of a reverse-biased silicon photodiode, the output current follows the intensity variations. This property makes silicon photodiodes useful for receiving modulated-light or modulated-IR signals.

> **Did You Know?**
> Silicon photodiodes exhibit greater sensitivity to radiant energy at some wavelengths than at others. The greatest sensitivity occurs in the *near infrared* part of the spectrum, at wavelengths just a little longer than the wavelength of visible red light.

Optoisolators

An LED or IRED and a photodiode can be combined in a single package to construct a component called an *optoisolator*. This device, shown schematically in Fig. 6-11, creates a modulated-light or modulated-infrared signal and sends it over a clear gap to a photodiode. An LED or IRED converts the electrical input signal to visible

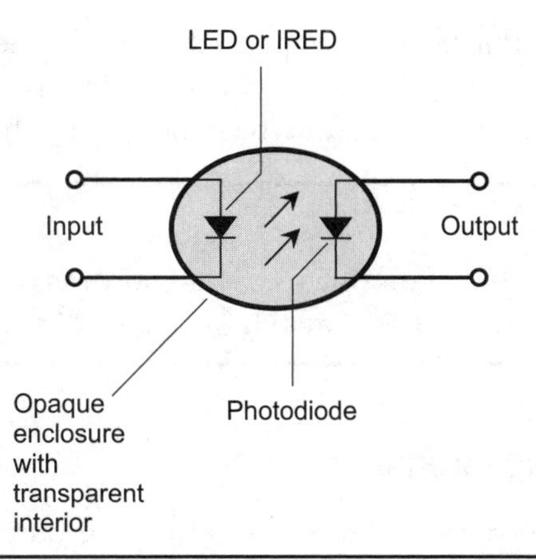

FIGURE 6-11 An optoisolator contains an LED or IRED at the input and a photodiode at the output with a transparent, electrically insulating medium, such as clear plastic or glass between them.

light or IR. The photodiode changes the visible light or IR back into an electrical signal, which appears at the output. An opaque enclosure prevents external light or IR from reaching the photodiode, ensuring that it receives only the radiation from the internal source.

When you want to transfer, or *couple*, an RF signal from one circuit to another, the two stages inevitably interact, to some extent, if you make the transfer by electrical means. The input impedance of a given stage, such as an amplifier, can affect the behavior of the circuits that feed power to it, leading to problems. Optoisolators overcome this effect because the coupling involves no electrical or magnetic effects. If the input impedance of the second circuit changes, the first circuit "sees" no change in the output impedance.

Photovoltaic Cells

A silicon diode with no applied bias can generate DC if strong enough IR, visible light, or ultraviolet (UV) radiation strikes its P-N junction. Scientists call this phenomenon the *photovoltaic effect*. Solar cells all work this way.

Photovoltaic (PV) *cells* are specially manufactured to have the greatest possible P-N junction surface area, maximizing the amount of radiant energy that strikes the junction. A single silicon PV cell can produce about 0.6 V of DC electricity in full daylight. If the illumination remains constant, the amount of current that a PV cell can deliver, and therefore, the amount of power that it can provide, depends on the surface area of its P-N junction.

You can connect multiple PV cells in series-parallel combinations to provide power for solid-state electronic devices, such as portable radios. The DC from these arrays can charge batteries, allowing for use of the electronic devices when radiant energy is not available (for example, at night or on dark days). A large assembly of solar cells, connected in series-parallel, is called a *solar panel*.

The power produced by a solar panel depends on the power from each individual PV cell, the number of cells in the panel, the intensity of the radiant energy that strikes the panel, and the angle at which the rays hit the surface of the panel.

Did You Know?

Some solar panels can produce several kilowatts of electrical power when the midday sun's unobstructed rays strike perpendicular to the panel surface.

Experiment 1: Voltage Reducer

You can use one or more diodes to reduce the output voltage of a DC battery in incremental amounts. You'll need two silicon rectifier diodes rated at 1 A and 600 PIV, available from Radio Shack as part number 276-1104. You'll also need 1/2-W resistors of 220, 330, 470, 680, 1000, 1500, and 3300 ohms.

How Does It Work?

As you know, it takes a certain minimum voltage to make current flow through a forward-biased semiconductor diode: the forward breaker voltage. In most diodes, it's a fraction of a volt, but it can vary slightly depending on how much current the diode carries.

If the forward-bias voltage across the diode P-N junction remains below the forward breaker voltage, the diode will not conduct. You can take advantage of this phenomenon in a surprising way! When you forward-bias a rectifier diode and connect it in series with a battery, the output voltage goes down to an extent equal to the forward breakover voltage.

Build and Test It

You can set up a voltage reducer with two diodes, as shown in Fig. 6-12, so that current flows through the load resistor R_L. Set your meter to indicate DC voltage. Connect the meter across the load resistance, paying attention to the polarity so that you'll get positive voltage readings. Try every resistor listed in Table 6-1 for R_L. Measure the voltage across R_L in each case. You'll have seven tests to do.

Tip

Whenever you do experiments involving resistors, measure their actual values before you connect them to anything else. After testing the resistors with my digital ohmmeter, I found that their values were 220, 326, 466, 671, 983, 1470, and 3240 ohms.

The load resistance affects the behavior of a diode-based voltage reducer. When you do these tests, you'll see that as the load resistance R_L goes down, the potential

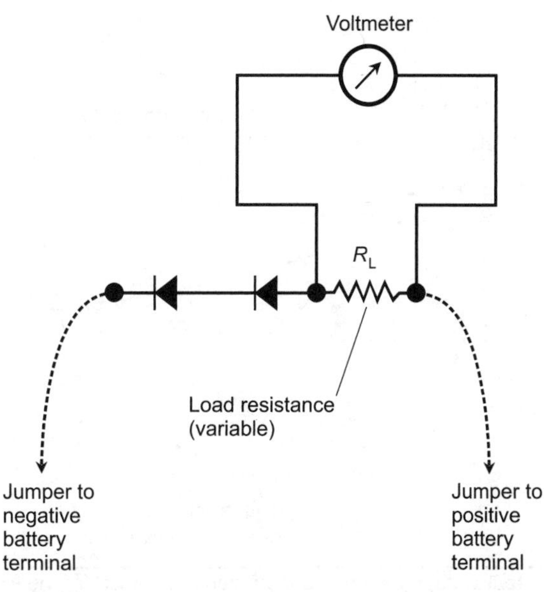

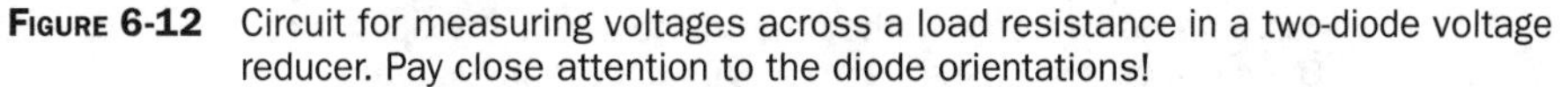

FIGURE 6-12 Circuit for measuring voltages across a load resistance in a two-diode voltage reducer. Pay close attention to the diode orientations!

difference (voltage drop) across it also goes down. In addition, the voltage across the load drops *more and more slowly* as R_L decreases at a constant rate. Table 6-1 shows the results that I got when I measured the voltages across various loads.

Plot your results as points on a coordinate grid with the load resistance on the horizontal axis and the voltage across the load resistor on the vertical axis, and then approximate the curve. When I did that, I got the graph of Fig. 6-13.

Repeat this experiment with only one diode. Then, if you like, get another package of diodes and try the experiment with three or four of them in series. You might also obtain some more resistors, covering a range of, say, 100 ohms to 100K ohms, and test the circuit using them for R_L.

TABLE 6-1 Outputs that I obtained with various loads connected to a diode-based voltage reducer. The circuit comprised two diodes rated at 1 A and 600 PIV, forward-biased and placed in series with a 6.30-V battery.

Load Resistance (ohms)	Output Voltage (volts)
3240	5.08
1470	4.99
983	4.96
671	4.91
466	4.88
326	4.84
220	4.79

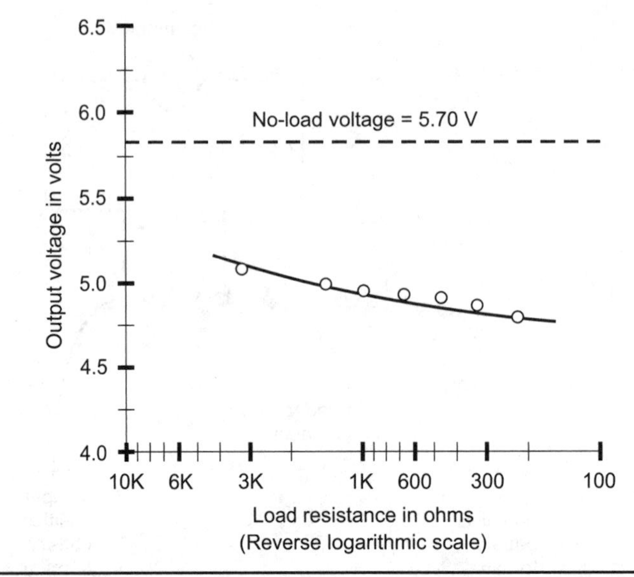

FIGURE 6-13 Output voltage as a function of load resistance for the voltage reducer. The solid curve shows my results. The dashed line shows the open-circuit (no-load) voltage. Open circles show measured voltages under various loads.

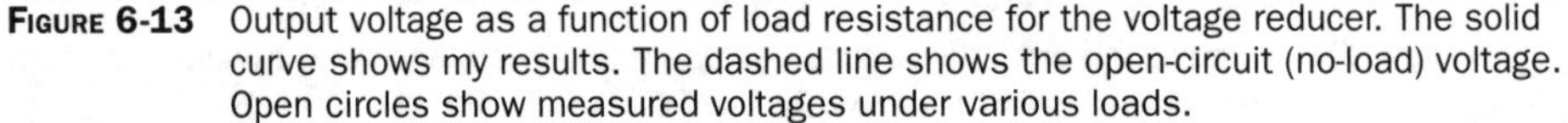

Don't use a resistor of less than about 75 ohms as the load. In this arrangement, a 1/2-W resistor of less than 75 ohms will let too much current flow, risking destruction of the resistor (and maybe the diodes too).

Experiment 2: Bridge Rectifier

In this experiment, you'll build a rectifier by connecting four diodes in a matrix called a *bridge*. Along with one of your Enercell transformers that you used in the last chapter's experiments, you'll need a power strip *without* a transient suppressor (or "surge protector"), four diodes rated at 3 A and 400 PIV, a resistor rated at 3300 ohms and 1/2 W, some jumpers, your digital meter, and a pair of rubber gloves.

How It Works

A *bridge rectifier* takes advantage of both halves of the AC cycle to produce pulsating DC, as shown in Fig. 6-14. Graph A shows the input AC waveform; graph B shows the rectified output. The entire wave gets through, but the bridge inverts every other half-cycle so that both half-cycles produce output of the same polarity.

With a full-wave bridge rectifier, the *effective* (*root-mean-square* or RMS) voltages of the input AC and the output DC are theoretically identical. In a practical circuit, because of the small voltage drops across the diodes, the RMS pulsating DC output voltage is slightly lower than the RMS AC input voltage.

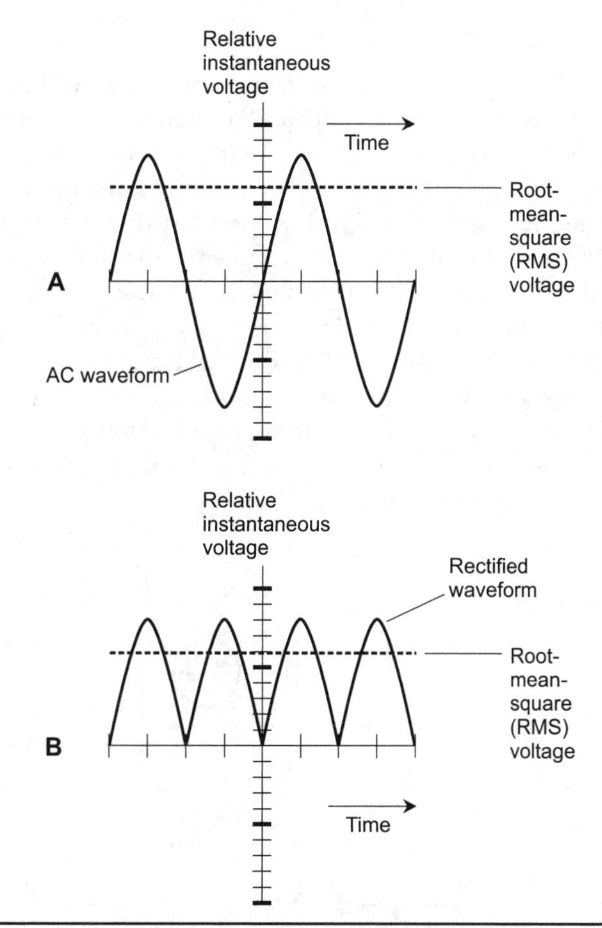

FIGURE 6-14 At A, an AC wave as it appears coming from a utility outlet. At B, the pulsating DC waveform as it appears at the output of a bridge rectifier.

Set the transformer to its lower-voltage output ("18 V"), plug it into the power strip, and switch the strip on. Measure the AC voltage at the transformer output. My meter indicated an AC output voltage of 20.8 V RMS. Yours will probably differ a little from mine, and it will also vary slightly from hour to hour as the load on your local utility fluctuates.

Did You Know?

The RMS voltage of an AC wave is the voltage that a DC source, such as a battery, would have to produce in order to heat up a load resistor (containing no inductance or capacitance) to the same extent as the AC wave does in the same amount of time. You might just as well call the RMS voltage the *DC-equivalent voltage*.

Build and Test It

Switch off the power strip. Connect the components, as shown in Fig. 6-15. Pay attention to the polarities of the diodes. Once you have all the components hooked up properly, switch the power strip on. Set the digital meter to indicate DC voltage. Hold the meter probe tips to the diode terminals at the points shown in Fig. 6-15. Note the meter reading. When I did this test, my meter displayed an open-circuit pulsating DC output from the bridge rectifier of 18.6 V RMS, which was 2.2 V less than the RMS AC output voltage from the transformer alone.

For a few moments, I wondered why the voltage difference was so great between the RMS AC input and the RMS pulsating DC output. I expected the difference to be somewhere between about 0.5 V and 1.2 V, but not 2.2 V! Then I looked closely at the component arrangements, and I saw that each half of the wave cycle must pass through two diodes in series. Therefore, the voltage gets "bumped down" twice with every DC pulse.

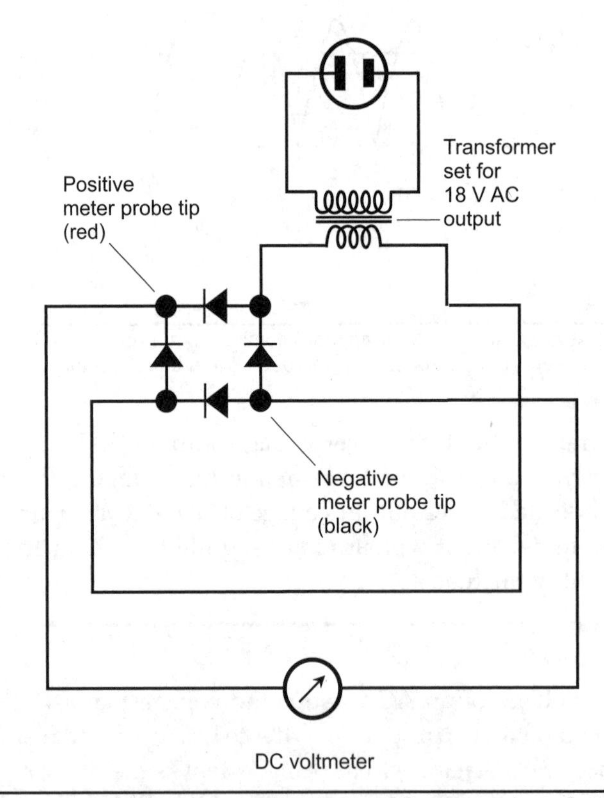

Figure 6-15 Measurement of the open-circuit pulsating DC output voltage from the bridge rectifier.

Output Voltage under Load

Switch off the power strip, and then connect one end of a 3300-ohm resistor to the point shown in Fig. 6-16. In this arrangement, the resistor will serve as a load for the rectifier output when you connect a jumper between the resistor's free end and the upper-left-hand corner of the bridge matrix.

If you use a different 3300-ohm resistor from the one you used in Experiment 1, measure the actual value of the resistor before doing any tests here. I used the same resistor, so I already knew that it had an actual value of 3240 ohms.

Switch on the power strip. Connect a jumper between the free end of the resistor and the point on the bridge matrix indicated in Fig. 6-16. Set your meter for DC volts. Place the meter probe tips at the locations shown to read the RMS pulsating DC output voltage across the resistor. My meter displayed 17.7 V, which was 0.9 V less than the open-circuit voltage.

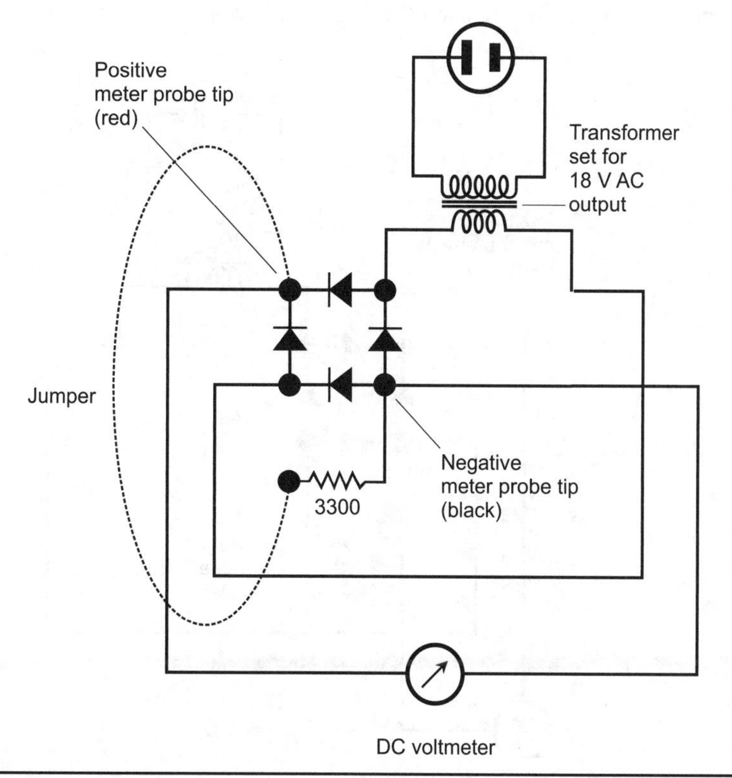

FIGURE 6-16 Measurement of the pulsating voltage from the bridge rectifier across a resistive load.

> **Tip**
>
> Don't be surprised if the voltage under load drops for you, too. Loads always drop the voltage, at least a little, at the output of a rectifier. The heavier the load (the lower the resistance), the more voltage drop you should expect.

Output Current through Load

You can use Ohm's Law to calculate the theoretical current that the full-wave bridge rectifier should drive through the load resistor. I measured a voltage of $E = 17.7$ V and a resistance of $R = 3240$ ohms, so I expected the current I to be

$$\begin{aligned} I &= E/R \\ &= 17.7/3240 \\ &= 0.00546 \text{ A} \\ &= 5.46 \text{ mA} \end{aligned}$$

Switch off the power strip. Disconnect the meter and set it for DC milliamperes. Connect the meter as shown in Fig. 6-17, and then switch the power strip

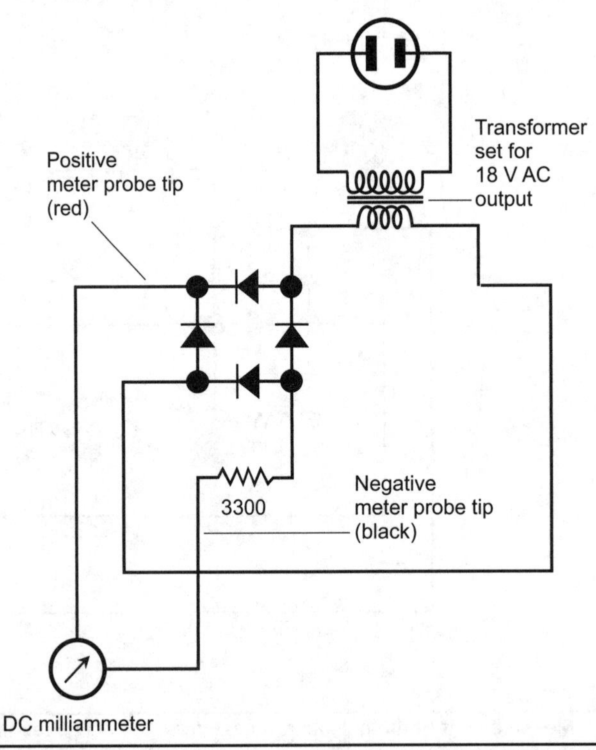

FIGURE 6-17 Measurement of the pulsating current that the bridge rectifier drives through a resistive load.

back on. What does your meter say? I measured 5.46 mA, the exact value that I predicted!

Variations on a Theme

Conduct the above-described calculations and measurements with a 1500-ohm resistor as the load. Then do the same maneuvers with two, three, and finally four 1500-ohm resistors in parallel. Using your ohmmeter, measure the value of each resistor combination ahead of time with the circuit powered-down. How does the rectifier output voltage change as the load resistance goes down? How does your calculated load current compare with the actual load current in each situation?

CHAPTER 7
Transistors

Transistors can generate, amplify, modulate, and mix AF and RF signals. Some transistors can also act as high-speed switches for DC and digital applications.

Bipolar Transistors

Bipolar transistors contain three sections of semiconductor material with two P-N junctions. Two main types prevail: a P type layer between two N type layers (called an *NPN transistor*), and an N type layer between two P type layers (called a *PNP transistor*).

NPN versus PNP

Figure 7-1A is a functional drawing of an NPN transistor, and Fig. 7-1B shows the schematic symbol. The P type, or center, layer forms the *base*. One of the N layers forms the *emitter*, and the other N layer forms the *collector*. The base, the emitter, and the collector are abbreviated as B, E, and C.

A PNP transistor has two P type layers, one on either side of a thin N type layer, as shown in Fig. 7-1C. The schematic symbol for this device appears in Fig. 7-1D. The N layer forms the base. One of the P layers forms the emitter, and the other P layer forms the collector.

A schematic will tell you whether the circuit designer intends a transistor to be NPN or PNP. The arrow always goes with the emitter, so you can identify the three electrodes without having to label them. In an NPN transistor, the arrow at the emitter points outward. In a PNP transistor, the arrow points inward.

In practical circuits, PNP and NPN transistors can perform the same functions. However, they require opposite voltage polarities, and the currents flow in the opposite directions. Often you can replace an NPN device with a PNP device or vice versa, reverse the power-supply polarity, and the new circuit will behave as the old one did.

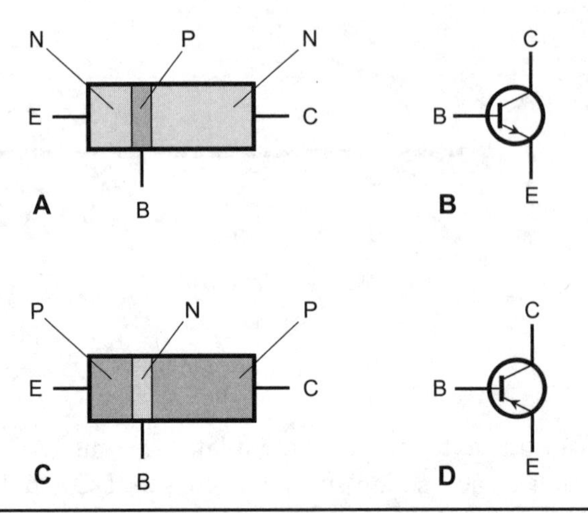

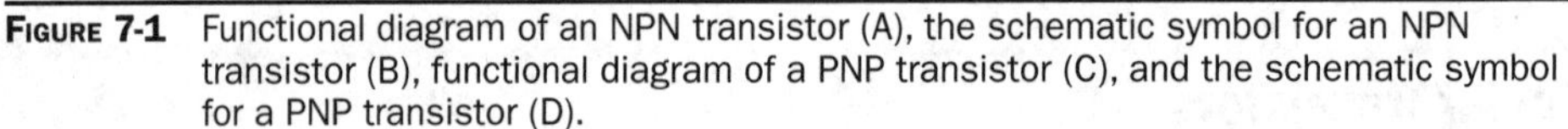

FIGURE 7-1 Functional diagram of an NPN transistor (A), the schematic symbol for an NPN transistor (B), functional diagram of a PNP transistor (C), and the schematic symbol for a PNP transistor (D).

> **Tip**
>
> Some bipolar transistors work best in RF amplifiers and oscillators; others work better for AF applications. Some can handle high power, and others are meant to deal with weak signals. Some are manufactured for digital operations, such as on/off switching and logic functions, instead of analog operations, such as oscillation and amplification.

Power Supply for NPN

In order to make a transistor work, you must apply certain voltages called *bias voltages* (or simply *bias*). In an NPN transistor, you'll usually connect the emitter to the negative battery or power-supply terminal and the collector to the positive terminal, often through coils and/or resistors. Figure 7-2 shows the basic emitter-

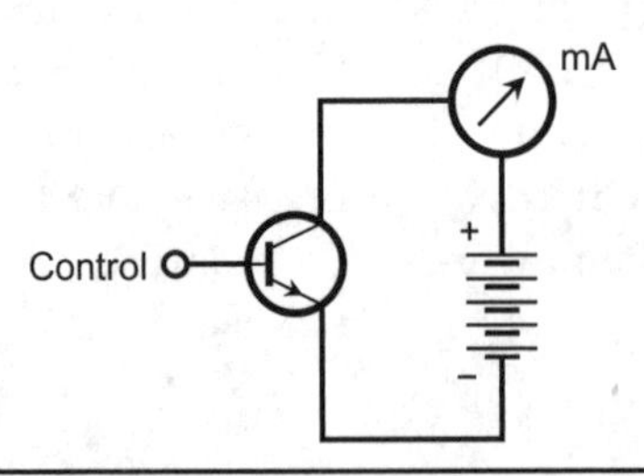

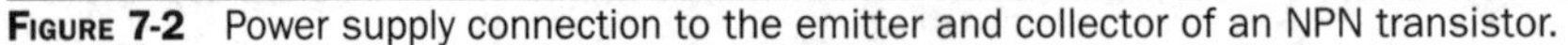

FIGURE 7-2 Power supply connection to the emitter and collector of an NPN transistor.

collector biasing scheme for an NPN transistor. Typical battery or power-supply voltages range from 3 V to 50 V.

In Fig. 7-2, you can label the base "Control" because, with constant collector-to-emitter (C-E) voltage (symbolized E_C or V_C), the magnitude of the current through the transistor depends on the current that flows across the emitter-base (E-B) junction, called the *base current* and labeled I_B. What happens at the base, from instant to instant in time, dictates how the device behaves as a whole.

Zero Bias

When you don't connect a bipolar transistor's base to anything, or when you short-circuit it to the emitter for DC, the transistor operates at zero *base bias* (or *zero bias*). No appreciable current can flow between the emitter and the collector with zero bias under no-signal conditions. In order to cause current to flow between the emitter and collector, you must apply a DC voltage equal to or greater than the forward breakover voltage at the E-B junction, or else introduce an AC signal at the base or emitter that causes the voltage at the E-B junction to reach or exceed the forward-breakover voltage for at least a small part of each signal cycle.

Reverse Bias

In the situation of Fig. 7-2, imagine that you connect a second battery between the base and the emitter of the NPN transistor, forcing the base to acquire a negative voltage with respect to the emitter. The addition of this new battery will cause the E-B junction to operate in a condition of *reverse bias*. (Assume that this new battery doesn't have a high enough voltage to cause avalanche breakdown at the E-B junction.)

When you reverse-bias the E-B junction of a transistor, no current flows between the emitter and the collector under no-signal conditions. You might inject a signal at the base or emitter to overcome the combined reverse-bias battery voltage and forward breakover voltage of the E-B junction, but such a signal must have positive voltage peaks high enough to cause conduction at the E-B junction for part of the input signal cycle. Otherwise, the transistor will remain in a state of *cutoff* for the entire cycle.

Forward Bias

Now suppose that, in the situation of Fig. 7-2, you make the bias voltage at the base of the NPN transistor positive relative to the emitter, starting at small levels and gradually increasing. You might do this by connecting a resistor between the base and the existing positive battery terminal, and another resistor between the base and the negative battery terminal. You vary the base bias by "tweaking" the values of the two resistors, which form a circuit called a *voltage divider*.

Alternatively, you can get base bias with a second battery. This action causes forward bias at the E-B junction. If this bias remains smaller than the forward breakover voltage, no current flows. But once the DC voltage reaches or exceeds the forward breakover threshold, the E-B junction conducts under no-signal conditions.

Despite a normal condition of reverse bias at the base-collector (B-C) junction, some *emitter-collector current*, more often called *collector current* and denoted I_C, flows when the E-B junction conducts. A small instantaneous rise in the *positive* polarity of an AC signal at the base, attended by a small instantaneous rise in the base current I_B, will cause a large instantaneous increase in the collector current I_C. Conversely, a small instantaneous rise in the *negative* polarity of an AC signal at the base, attended by a small instantaneous drop in I_B, will cause a large instantaneous decrease in I_C. The collector current varies more than the base current does, so the transistor acts as an *alternating-current (AC) amplifier*.

Saturation

If you adjust the conditions at the E-B junction so that I_B continues to rise, you'll eventually reach a point where I_C no longer increases rapidly. Ultimately, the I_C versus I_B function, or *characteristic curve*, levels off. Figure 7-3 shows a *family of characteristic curves* for a typical bipolar transistor (either NPN or PNP; the graph is not polarity-specific). Each curve shows the situation for a certain fixed *collector-to-emitter voltage*, usually called the *collector voltage* and symbolized as E_C. The actual current levels depend on the internal structure of the transistor. Where the curves level off, the transistor operates in a state of *saturation*. Under these conditions, it cannot amplify, but it can operate as a switch if you change back and forth between the saturated state (switch on) and the cutoff state (switch off).

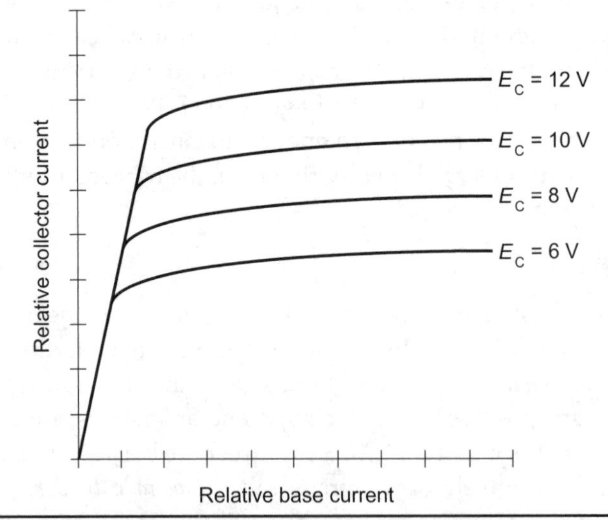

FIGURE 7-3 A family of characteristic curves for a bipolar transistor.

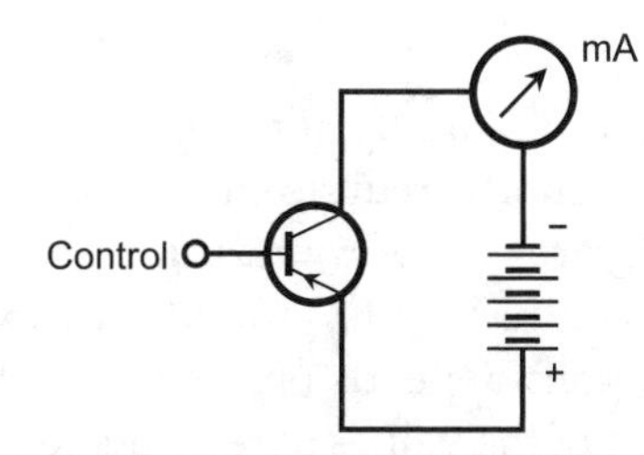

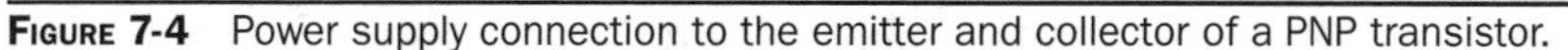

Figure 7-4 Power supply connection to the emitter and collector of a PNP transistor.

> **Tip**
>
> When working with analog circuits such as amplifiers, engineers rarely operate bipolar transistors in the saturated state. However, they often bias bipolar transistors at saturation in digital circuits in which the signal is always either full-on (high, or logic 1) or full-off (low, or logic 0).

Power Supply for PNP

For a PNP transistor, the DC power supply connection to the emitter and collector is a mirror image of the case for an NPN device, as shown in Fig. 7-4. You reverse the battery polarity compared with the NPN circuit. To overcome forward breakover at the E-B junction, an applied voltage or signal at the base (labeled "Control") must attain sufficient negative peak polarity. If you apply a positive DC voltage at the base, the device will operate in a state of cutoff.

Either type of transistor, PNP or NPN, works as a *current valve*. Small changes in the base current I_B induce large fluctuations in the collector current I_C when you operate the device in the region of the characteristic curve in which the graph has a steep rise over run (technically called *slope*). The details of the internal atomic activity differ in the PNP device as compared with the NPN device, but in most practical cases, the external circuitry can't tell the difference.

Basic Bipolar-Transistor Circuits

You can connect a bipolar transistor to external components to build circuits for specialized tasks. Three general arrangements prevail:

1. The *common-emitter circuit*, in which you ground the emitter for signal
2. The *common-base circuit*, in which you ground the base for signal
3. The *common-collector circuit*, in which you ground the collector for signal

Common Emitter

Figure 7-5 is a schematic diagram of a generic NPN common-emitter circuit. Capacitor C_1 presents a short circuit to the AC signal, placing the emitter at *signal ground* (but not necessarily DC ground). Resistor R_1 gives the emitter a small positive DC voltage with respect to ground. The exact DC voltage at the emitter under no-signal conditions depends on the value of R_1 and also on the base bias, determined by the ratio of the values of resistors R_2 and R_3. The base bias can range from 0 V (ground potential) all the way up to the supply voltage, in this case +12 V DC. Normally, you'll want to set the no-signal base bias a couple of volts positive with respect to DC ground.

Capacitor C_2 isolates, or *blocks*, DC from the input, while allowing the AC signal to get through. Capacitor C_3 blocks DC at the output, while letting the AC signal pass. Engineers call capacitors, such as C_2 and C_3, *blocking capacitors*. Resistor R_4 keeps the output signal from shorting through the power supply, while, nevertheless, allowing a positive DC voltage to exist at the collector.

A signal enters the circuit through C_2, where it causes the base current I_B to vary. Small fluctuations in I_B cause large variations in I_C. This current passes through R_4, causing a fluctuating DC voltage to appear across that resistor. The AC signal, superimposed on the DC flowing through R_4, passes through C_3 to the output terminal.

The common-emitter configuration offers excellent gain when properly designed to operate as an amplifier. The AC output wave appears in *phase opposition* to the

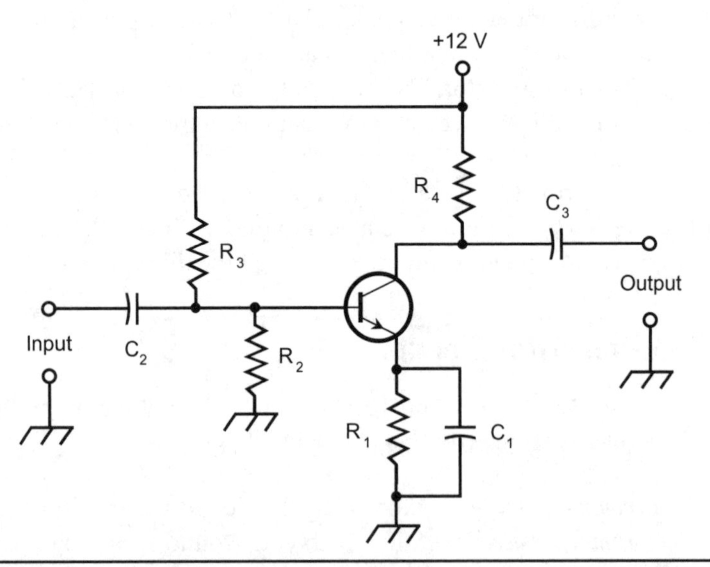

FIGURE 7-5 Common-emitter bipolar-transistor configuration.

input wave if you input a sine wave to the circuit. If circuitry outside the transistor inverts the signal again, a high-gain common-emitter circuit will sometimes oscillate rather than amplify.

Tip

You can minimize the risk of oscillation in a common-emitter amplifier by making sure that the circuit doesn't have too much gain. (An amplifier can sometimes work too well for its own good!) Alternatively, you can use a different circuit configuration, such as the common-base design described next.

Common Base

In the common-base circuit (Fig. 7-6), you place the base at signal ground. The DC bias on the transistor is the same for this circuit as for the common-emitter circuit. You apply the AC input signal to the emitter, producing fluctuations in the voltage across R_1, in turn causing small variations in I_B. As a result, you get a large change in I_C, the collector current that flows through R_4, so amplification occurs. The output wave appears in phase coincidence (lockstep) with the input wave.

The signal enters through C_1. Resistor R_1 keeps the input signal from shorting to ground. Base bias is provided by R_2 and R_3. Capacitor C_2 keeps the base at signal ground. Resistor R_4 keeps the signal from shorting through the power supply. The AC part of the output signal goes through C_3.

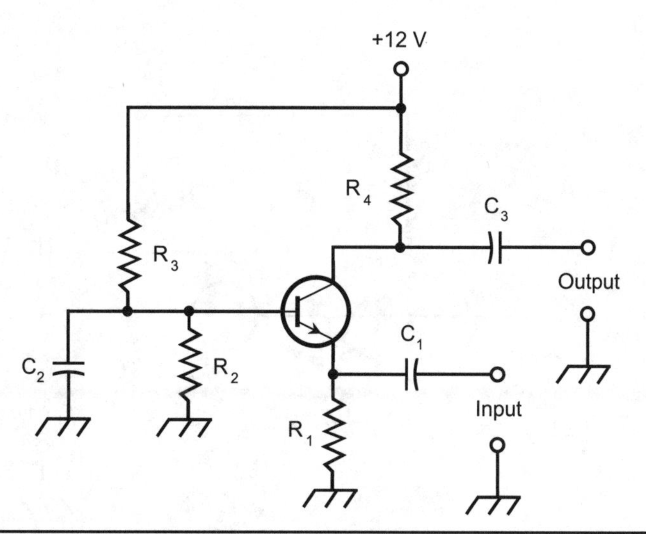

Figure 7-6 Common-base bipolar-transistor configuration.

A common-base circuit can't produce as much gain as a common-emitter circuit does, but the common-base amplifier is less prone to oscillate because it's less susceptible to the undesirable effects of *positive feedback,* which can cause an amplifier to "get out of control" in much the same way as a public-address system howls or rumbles when the microphone picks up too much sound from the speakers.

> **Tip**
>
> Common-base circuits work well as high-power amplifiers in radio transmitters, where too much positive feedback can cause *parasitic oscillation*—so called because it robs the transmitter of useful signal output power on its design frequency. "Parasitics" can wreak havoc on the airwaves by interfering with other wireless communications.

Common Collector

A common-collector circuit (Fig. 7-7) operates with the collector at signal ground. The input signal passes through C_2 onto the base of the transistor. Resistors R_2 and R_3 provide the base bias. Resistor R_4 limits the current through the transistor. Capacitor C_3 keeps the collector at signal ground. A fluctuating current flows through R_1, and a fluctuating voltage, therefore, appears across it. The AC component

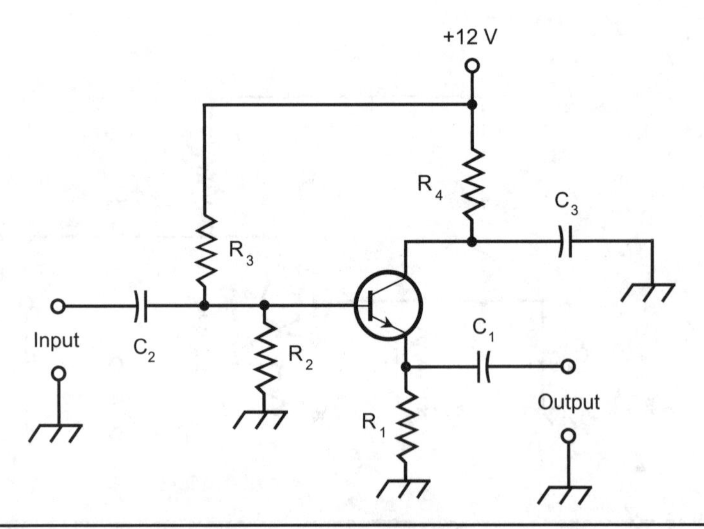

FIGURE 7-7 Common-collector bipolar-transistor configuration.

passes through C_1 to the output. Because the instantaneous output signal level follows along with the instantaneous emitter current, this circuit is sometimes called an *emitter follower*. The output wave appears in phase with the input wave. When well designed, an emitter follower works over a wide range of frequencies, and offers a low-cost alternative to an RF transformer.

Did You Know?

An emitter-follower circuit won't amplify signals, but it can help to provide isolation between two different parts of an electronic system. Engineers call this sort of isolation circuit a *buffer*.

Tip

Figures 7-5, 7-6, and 7-7 show NPN transistor circuits. You can obtain the equivalent diagrams for PNP circuits by replacing the NPN transistors with PNP devices, and by reversing the power-supply polarity in each case (providing about –12 V DC, rather than +12 V DC, at the non-collector end of resistor R_4).

Field-Effect Transistors

The other major form of semiconductor transistor (besides the bipolar type) is the *field-effect transistor* (FET). Two main versions exist: the *junction FET* (JFET) and the *metal-oxide-semiconductor FET* (MOSFET).

Principle of the JFET

In a JFET, a fluctuating electric field causes the current to vary within the semiconductor medium. Charge carriers (electrons or holes) move along a path called the *channel* from the *source* (S) electrode to the *drain* (D) electrode. As a result, you observe a drain current I_D equal to the source current I_S, and also equal to the current at any point along the channel. The current through the channel depends on the instantaneous voltage at the *gate* (G) electrode.

If you design a JFET circuit properly, then small changes in the gate voltage, E_G, cause large changes in the current through the channel, and therefore, in I_D. When the fluctuating drain current passes through an external resistance, you get large variations in the instantaneous DC voltage across that resistance. You can "draw off" the AC part of the fluctuating DC, thereby obtaining an output signal much stronger than the input signal. That's how an FET produces *voltage amplification*.

N-Channel versus P-Channel

Figure 7-8A is a simplified functional drawing of an *N-channel JFET*. Figure 7-8B shows its schematic symbol. The N type material forms the path for the current. Electrons constitute most of the charge carriers, so by definition, electrons are the *majority carriers*. The drain is connected to the positive power-supply terminal, usually through a resistor, a coil, or some other combination of components. The gate comprises P type material. Another, larger section of P type material, called the *substrate*, forms a boundary on the side of the channel opposite the gate and serves as a "foundation" for the JFET physical structure. The voltage on the gate produces an electric field that interferes with the flow of charge carriers through the channel. As E_G becomes more negative, the electric field chokes off the current though the channel more and more, so the drain current I_D decreases.

A *P-channel JFET* (Figs. 7-8C and D) has a current pathway of P type semiconductor material. The majority carriers are holes. The drain is connected to the negative power-supply terminal. The gate and substrate consist of N type material. The

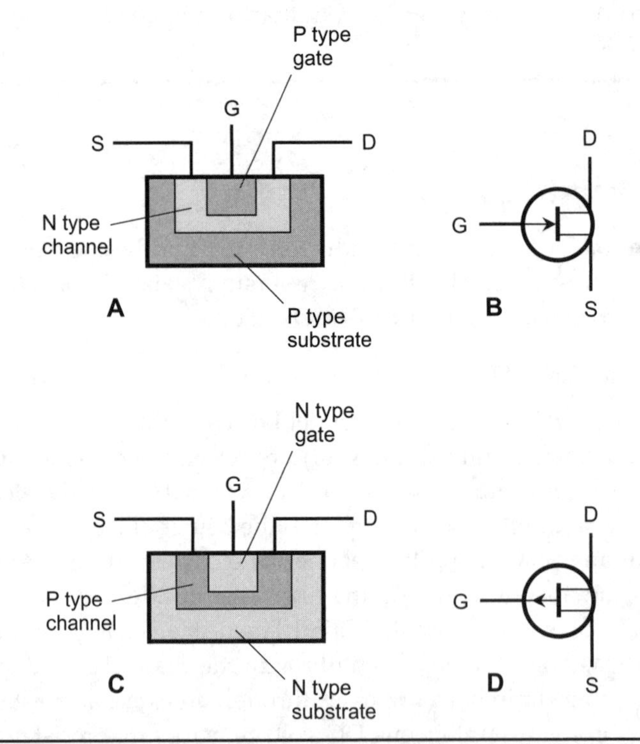

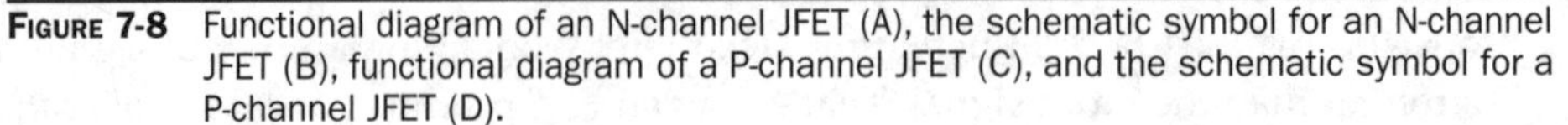

FIGURE 7-8 Functional diagram of an N-channel JFET (A), the schematic symbol for an N-channel JFET (B), functional diagram of a P-channel JFET (C), and the schematic symbol for a P-channel JFET (D).

more positive E_G gets, the more the electric field chokes off the current through the channel, and the smaller I_D becomes.

You can usually recognize an N-channel JFET in schematic diagrams by an arrow pointing inward at the gate, and a P-channel JFET by an arrow pointing outward at the gate. Some diagrams lack arrows in JFET symbols, but the power-supply polarity gives away the device type. When the drain goes to the positive power-supply voltage (with the negative power-supply terminal connected to ground), it indicates an N-channel JFET. When the drain goes to the negative power-supply voltage (with the positive power-supply terminal connected to ground), it indicates a P-channel JFET.

Tip

If you replace an N-channel JFET with a P-channel JFET and reverse the power-supply polarity, in most cases the new circuit will operate pretty much the same way as the old one did, as long as the new JFET has the proper specifications.

Depletion and Pinchoff

A JFET works because the voltage at the gate produces an electric field that interferes, more or less, with the flow of charge carriers along the channel.

As the drain voltage E_D increases, so does the drain current I_D (up to a certain maximum leveling-off value) as long as the gate voltage E_G remains constant, and as long as E_G doesn't get too high. As E_G increases (negatively in an N channel or positively in a P channel), a *depletion region* forms in the channel. Charge carriers can't flow in the depletion region, so they must pass through a narrowed channel. Because of the restricted pathway, the current goes down.

As the gate voltage E_G increases (negatively for an N-channel device or positively for a P-channel device), the depletion region widens and the channel narrows. If E_G gets high enough, the depletion region closes the channel, preventing any flow of charge carriers from the source to the drain. Engineers call this condition *pinchoff*.

JFET Biasing

Figure 7-9 shows two biasing arrangements for an N-channel JFET. At A, the gate is grounded through resistor R_2. The source resistor R_1 limits the current through the device. Resistors R_1 and R_2 determine the gate bias. The drain current I_D flows through R_3, producing a voltage across it. The AC output signal passes through C_2.

At B, the gate is connected through potentiometer R_2 to a negative DC voltage source. When you adjust this potentiometer, you vary the negative gate voltage E_G at the point between resistors R_2 and R_3. Resistor R_1 limits the current through

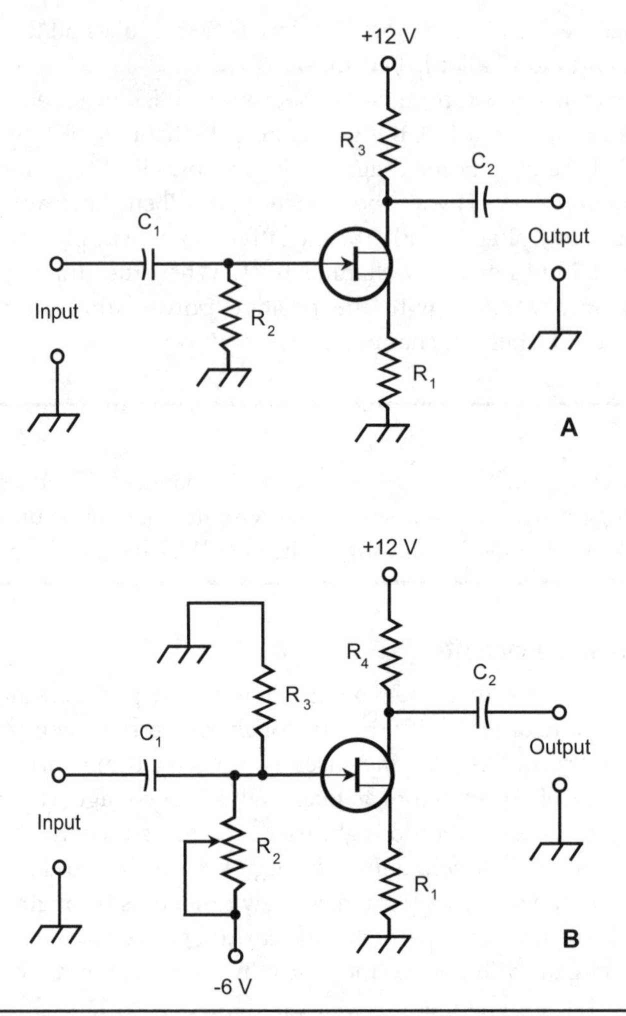

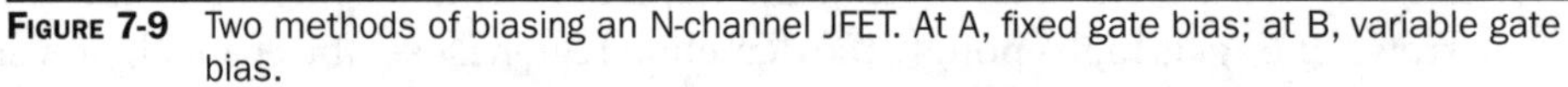

FIGURE 7-9 Two methods of biasing an N-channel JFET. At A, fixed gate bias; at B, variable gate bias.

the JFET. The drain current I_D flows through R_4, producing a voltage across it. The AC output signal passes through C_2.

In both of these circuits, you would connect the drain to a positive DC voltage source relative to ground. In the case of a P-channel JFET circuit, the polarities must be reversed, so you would connect the drain to a negative DC voltage source relative to ground.

The biasing arrangement in Fig. 7-9A is commonly used for weak-signal amplifiers, low-level amplifiers, and oscillators. You would more likely use the scheme shown in Fig. 7-9B in power amplifiers requiring a substantial input signal.

> **Tip**
>
> Typical JFET power-supply voltages are comparable to those with bipolar transistors. The voltage E_D between the drain and ground can range from approximately 3 V to 50 V; most often it's 6 V to 12 V.

How the JFET Amplifies

Figure 7-10 shows the relative drain current I_D as a function of the no-signal gate bias voltage E_G for a hypothetical N-channel JFET, assuming that the drain voltage E_D remains constant. When a JFET operates properly, the drain current is the same as the current through the channel, so you can think of the drain current as the channel current too.

When E_G is fairly large and negative, the JFET operates in a state of pinchoff, so no current flows through the channel. As E_G gets less negative, the channel opens up and current begins flowing. As E_G gets still less negative, the channel grows wider and the drain current I_D increases. As E_G approaches the point where the source-gate (S-G) junction reaches the forward breakover voltage, the channel conducts as well as it can; it's "wide open." If E_G gets still more positive, exceeding the forward breakover voltage and causing the S-G junction to conduct, some of the current in the channel leaks out through the gate. In a typical JFET circuit, S-G junction current should never flow.

The greatest amplification for weak signals occurs when you set E_G so that the curve in Fig. 7-10 has its steepest slope, as shown by the range marked *X*. In a

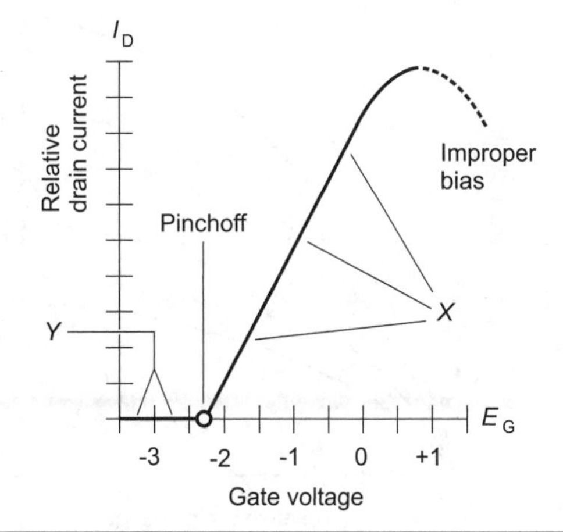

FIGURE 7-10 Relative drain current as a function of gate voltage for a hypothetical N-channel JFET.

high-power RF transmitting amplifier in which the input signal is relatively powerful to begin with, you'll often get the best results when you bias a JFET at or beyond pinchoff, in the range marked *Y*.

> **Tip**
>
> From precalculus, you might recall that in a rectangular coordinate graph, such as Fig. 7-10, the part of a curve that ramps up as you move toward the right has a *positive slope* by definition; the part of a curve that ramps down as you move toward the right has a *negative slope* by definition.

In a practical JFET amplifier circuit, the drain current passes through the drain resistor, as shown in Fig. 7-9A or Fig. 7-9B. Small fluctuations in E_G cause large changes in I_D, and these variations, in turn, produce wide swings in the DC voltage across R_3 (at A) or R_4 (at B). The AC (signal) component of this voltage goes through capacitor C_2, and appears at the output with greater peak-to-peak voltage than that of the input signal at the gate. Therefore, the JFET operates as a *voltage amplifier*.

Drain Current versus Drain Voltage

With any JFET, you can test and graph the drain current I_D as a function of the drain voltage E_D for various values of gate voltage E_G. The resulting graph is called a *family of characteristic curves* for the device. Fig. 7-11 shows a family of characteristic curves for a hypothetical N-channel JFET.

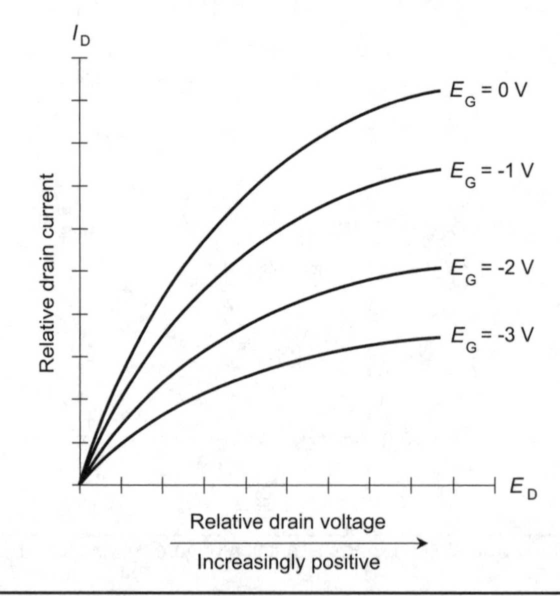

FIGURE 7-11 A family of characteristic curves for an N-channel JFET.

Metal-Oxide FETs

The acronym MOSFET (pronounced "MOSS-fet") stands for *metal-oxide-semiconductor field-effect transistor*. This type of component can be constructed with a channel of N type material, or with a channel of P type material. Engineers call the former type an *N-channel MOSFET* and the latter type a *P-channel MOSFET*. Figure 7-12A is a functional cross-section drawing of an N-channel MOSFET. Figure 7-12B shows the schematic symbol. The P-channel cross-section drawing and symbol appear at C and D.

The Insulated Gate

When semiconductor engineers conceived and developed the MOSFET, they called it an *insulated-gate FET* or IGFET. Some people think that this expression describes the device better than the currently accepted term does. The gate electrode is actually insulated, by a thin layer of dielectric material, from the channel.

The input impedance for a MOSFET exceeds that of a JFET when you apply an input signal at the gate electrode. In fact, the gate-to-source (G-S) resistance of a

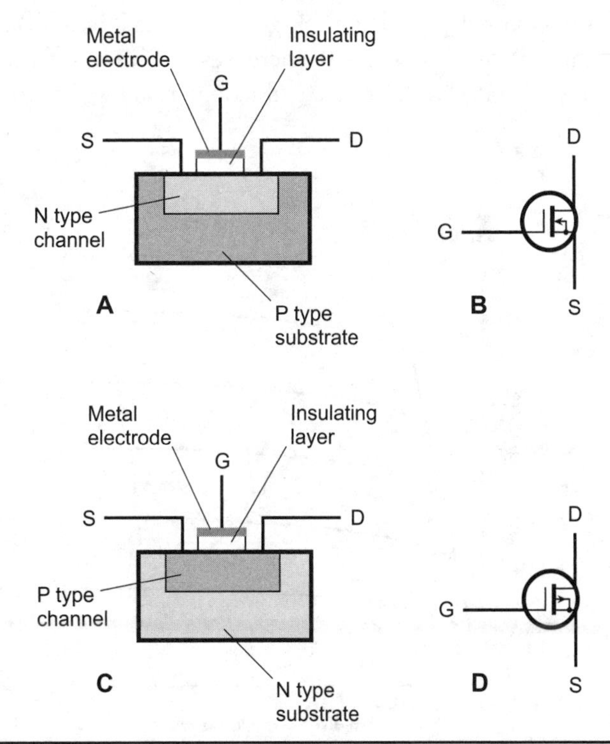

Figure 7-12 Pictorial diagram of an N-channel MOSFET (A), the schematic symbol for an N-channel MOSFET (B), pictorial diagram of a P-channel MOSFET (C), and the schematic symbol for a P-channel MOSFET (D).

typical MOSFET compares favorably to the *leakage resistance* of a capacitor. It's so large that you can consider it infinite in most applications.

Figure 7-13 shows a family of characteristic curves for a hypothetical N-channel MOSFET. Note that the curves rise steeply at first for relatively small values of drain voltage E_D, but as E_D increases beyond a certain threshold, the curves level off more quickly than they do for a JFET.

Beware and Take Care!

Metal-oxide devices are easily destroyed by electrostatic discharges, even small ones. When you build, test, or service a circuit containing MOS transistors or integrated circuits, you must use special equipment to ensure that your hands don't acquire any electrostatic charge. If any stray discharge occurs through the thin, fragile dielectric layer in a MOSFET, the resulting current can destroy the device. I've seen it occur even in the humid summer climate of South Florida. So beware and take care!

Depletion versus Enhancement Modes

In a JFET, the channel conducts with zero gate bias, when the gate has the same voltage as the source ($E_G = 0$). As E_G increases (negatively for an N-channel device and positively for a P-channel device), the depletion region grows wider and wider,

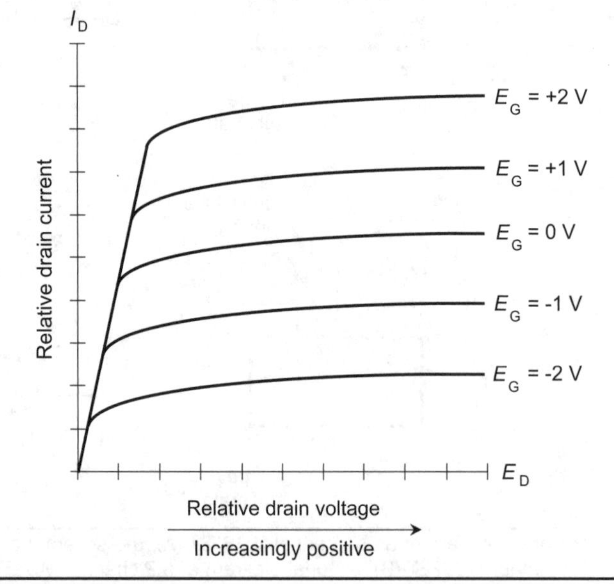

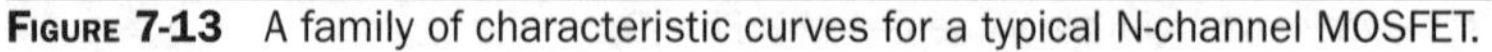

FIGURE 7-13 A family of characteristic curves for a typical N-channel MOSFET.

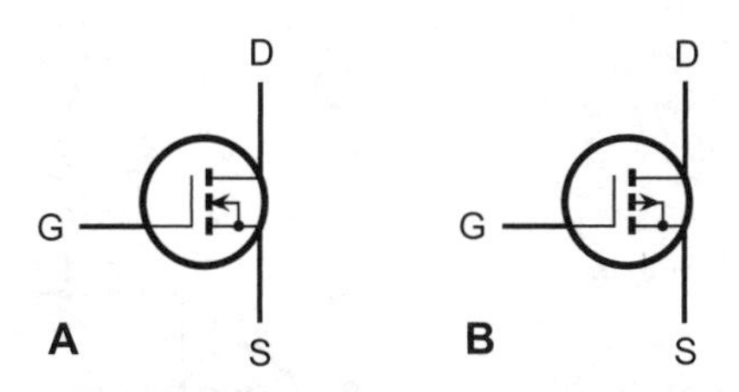

Figure 7-14 At A, the schematic symbol for an N-channel enhancement-mode MOSFET. At B, the schematic symbol for a P-channel enhancement-mode MOSFET.

so the charge carriers must pass through a narrower and narrower channel. This condition is known as the *depletion mode.* Some MOSFETs can function in the depletion mode. The drawings and schematic symbols of Fig. 7-12 depict the internal construction and schematic symbols for *depletion-mode* MOSFETs.

Metal-oxide-semiconductor technology allows for an alternative electrical environment that radically differs from the depletion mode. An *enhancement-mode* MOSFET has a pinched-off channel at zero bias. You must apply a gate bias voltage, E_G, to create a channel in this type of device. If $E_G = 0$, then $I_D = 0$ in the absence of signal input. You apply gate bias and signals to widen, rather than constrict, the channel.

Figure 7-14 shows the schematic symbols for N-channel and P-channel enhancement-mode MOSFETs. Note that in these symbols, the right-hand vertical lines are broken, rather than solid as in the symbols for depletion-mode MOSFETs. That difference allows you to distinguish between the two types of device when you see them in circuit diagrams.

Basic FET Circuits

Three general circuit configurations exist for FETs. They're the equivalents of the common-emitter, common-base, and common-collector bipolar-transistor circuits, and they break down as follows:

1. The *common-source circuit,* in which you ground the source for AC signals.
2. The *common-gate circuit,* in which you ground the gate for AC signals.
3. The *common-drain circuit,* in which you ground the drain for AC signals.

Common Source

In a *common-source circuit,* you apply the input signal to the gate, as shown in Fig. 7-15. This diagram shows an N-channel JFET, but you could substitute an N-channel, depletion-mode MOSFET and get the same results in a practical circuit. You could also use an N-channel, enhancement-mode MOSFET and add an extra resistor between the gate and the positive power-supply terminal.

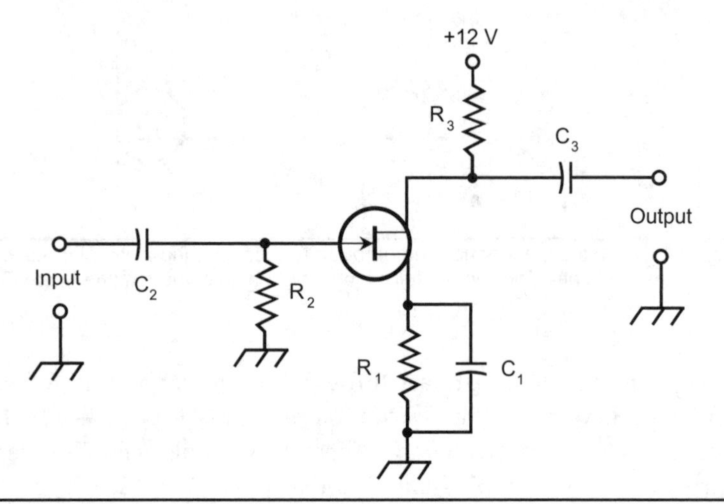

FIGURE 7-15 Common-source FET configuration.

For P-channel devices, the power supply would provide a negative voltage rather than a positive voltage. Otherwise, the circuit details would correspond to those shown in Fig. 7-15.

Capacitor C_1 and resistor R_1 place the source at signal ground, while elevating the source above ground for DC. The AC signal enters through capacitor C_2. Resistors R_1 and R_2 provide bias for the gate. The AC signal passes out of the circuit through capacitor C_3. Resistor R_3 keeps the output signal from shorting through the power supply.

The circuit of Fig. 7-15 can offer a starting point for the design of weak-signal amplifiers and low-power oscillators, especially in RF systems. The common-source arrangement provides the greatest gain of the three FET circuit configurations. The output wave is inverted (appears in phase opposition) with respect to the input wave.

Common Gate

The *common-gate circuit* (Fig. 7-16) operates with the gate at signal ground. You apply the input signal to the source. This illustration shows an N-channel JFET. For other types of FETs, the same considerations apply as in the case of the common-source circuit. An N-channel enhancement-mode device requires a resistor between the gate and the positive power-supply terminal. For P-channel devices, you reverse the polarity of the power supply.

The DC bias for the common-gate circuit resembles that for the common-source arrangement, but the signal follows a different path. The AC input signal enters through capacitor C_1. Resistor R_1 keeps the input from shorting to ground. Resistors R_1 and R_2 provide the gate bias. Capacitor C_2 places the gate at signal ground, while allowing DC bias voltage to exist on that electrode. The output signal exits

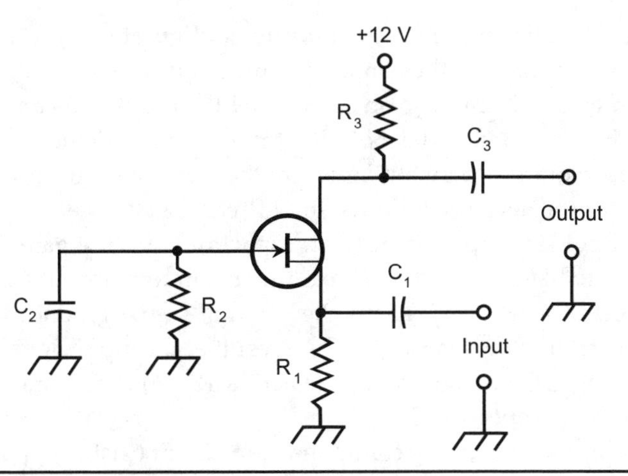

FIGURE 7-16 Common-gate FET configuration.

the circuit through C_3. Resistor R_3 keeps the output signal from shorting through the power supply.

The common-gate arrangement produces less gain than its common-source counterpart. However, a common-gate amplifier is less likely than a common-source amplifier to break into unwanted oscillation. The output wave occurs in phase coincidence with the input wave.

Common Drain

Figure 7-17 shows a *common-drain circuit,* in which you place the drain at signal ground. Engineers sometimes call this circuit a *source follower* because the output signal waveform follows the instantaneous voltage at the source electrode.

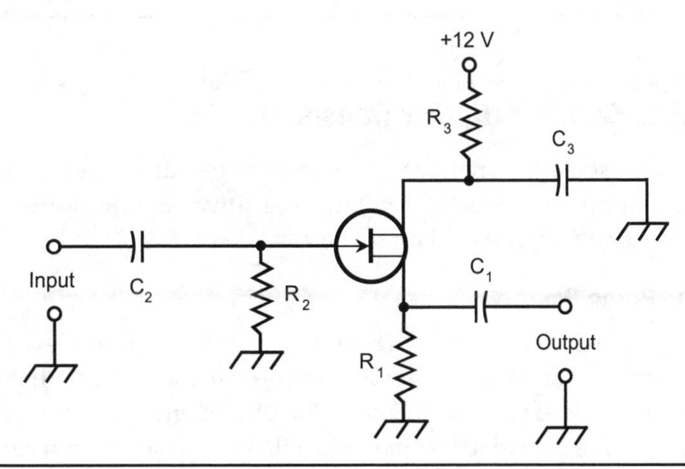

FIGURE 7-17 Common-drain FET configuration.

To use an FET in the common-drain configuration, you bias it for DC in the same way as you do in the common-source and common-gate circuits. The circuit shown in Fig. 7-17 employs an N-channel JFET, but you can substitute any other kind of FET, reversing the polarity for P-channel devices. Enhancement-mode MOSFETs require a resistor between the gate and the positive power-supply terminal (or the negative terminal for a P-channel device).

The input signal passes through capacitor C_2 to the gate. Resistors R_1 and R_2 provide gate bias. Resistor R_3 limits the maximum current that can flow through the channel. Capacitor C_3 keeps the drain at signal ground. Fluctuating DC (the channel current) flows through R_1 as a result of the input signal; this current causes a fluctuating DC voltage to appear across R_1. You take the AC output from the source through capacitor C_1.

The output wave of the common-drain circuit exists in phase coincidence with the input wave. This circuit, like the common-collector arrangement, can serve as a low-cost alternative to a wideband transformer, especially in RF applications.

Some Tech Talk

In the preceding discussions (accompanied by Figs. 7-15, 7-16, and 7-17), you might get the idea that you can directly interchange JFETs and MOSFETs in practical circuits. Sometimes you can do that, but not always. Although the circuit diagrams for JFET and depletion-mode MOSFET devices look identical when you don't specify actual component values, there's usually a difference in the optimum resistances and capacitances. These optimum values can vary not only between JFETs and MOSFETs, but also between depletion-mode MOSFETs and enhancement-mode MOSFETs. Specific design examples would surpass the scope of this book, but you should remember that JFETs and MOSFETs are sometimes, but not always, directly interchangeable.

Experiment 1: Check a Bipolar Transistor

You can use some (but not all) ohmmeters to test diodes, bipolar transistors, and JFETs by using the meter as a polarity-sensitive continuity tester. But you must make sure that your meter is up to the task.

Test the Probe Polarity

First, you must determine which ohmmeter probe produces what polarity. You might imagine that electrons would emerge from the probe with the black wire (the negative meter lead for current and voltage), and enter the probe with the red wire (the positive meter lead for current and voltage), so the electron current would flow from black to red. Maybe they do that in your meter. But when I tested my analog ohmmeter using a separate microvolt meter, and also using a couple of rectifier diodes known to be good, the opposite situation prevailed. Because of the way that particular ohmmeter's internal battery is connected, the red meter lead produces negative DC and the black one produces positive DC, so the electrons actually move from red to black!

Heads Up!

Not all multimeters are polarity-reversed for current production as opposed to measurement, but my analog meter is. I'm glad that I found out about it before starting the component tests. You should do the same thing.

Check the Probe Voltage

The other lurking bugaboo that might mess up this experiment involves the actual voltage produced at the test leads of your ohmmeter. If it's less than the forward breakover voltage of a typical P-N junction (about 0.6 V), your meter won't indicate conduction in either direction through the junction, and you might mistakenly conclude that the component is bad. My digital meter failed to produce enough voltage, and showed extremely high resistance in either direction through the P-N junctions in several different components. My analog meter did produce the necessary voltage, and I found that out by testing a rectifier diode known to be good. With the diode's anode positive and the cathode negative, my meter showed about 40 ohms. With the anode negative and the cathode positive, my meter showed "infinity" ohms.

Tip

Most diodes have a line on the case to indicate the cathode side. My rectifier diodes, having black cases, have white lines on the cathode sides.

Get a Transistor

You'll find NPN bipolar transistors at most Radio Shack retail outlets, and also on their website (www.radioshack.com). A small device will suffice; you don't need a big power transistor. I used a 2N222 transistor, Radio Shack catalog number 276-1617, but any similar component will work as well. Once you get the transistor, you'll probably decide to get a pair of strong reading glasses too. That transistor is small, and unless you have keen eyesight and a steady hand, you'll have trouble holding the meter probe tips up against the transistor leads. Worse yet, you might find it difficult to see the leads themselves.

Prepare for Testing

Before you test the transistor, make certain that you know which lead goes to the emitter, which one goes to the base, and which one goes to the collector. The wrapping package for the component should have a printed guide on the back, telling you how to locate them. Once you know that information, don't connect the transistor into a circuit. Test it "in isolation" first.

Go for It!

The tests involve holding the meter probe tips up against the transistor leads in all six ways that it can be done. Your body resistance is high enough so that you should not have to wear gloves. Figures 7-18 through 7-23 show the connections and indicate the results you should see with a good NPN transistor. Moderate resistance on the order of a few ohms or tens of ohms (the exact value doesn't matter) indicates conduction. "Infinite" resistance indicates non-conduction.

1. Electrons should flow from the emitter to the base (Fig. 7-18).
2. Electrons should not flow from the base to the emitter (Fig. 7-19).
3. Electrons should flow from the collector to the base (Fig. 7-20).
4. Electrons should not flow from the base to the collector (Fig. 7-21).
5. Electrons should not flow from the emitter to the collector (Fig. 7-22).
6. Electrons should not flow from the collector to the emitter (Fig. 7-23).

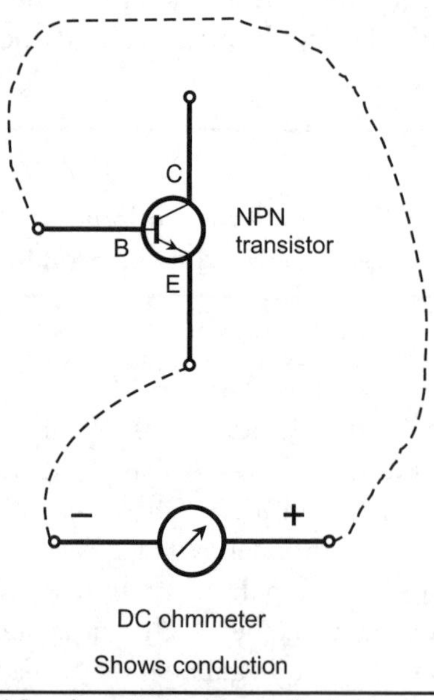

FIGURE 7-18 Electrons should flow from the emitter to the base.

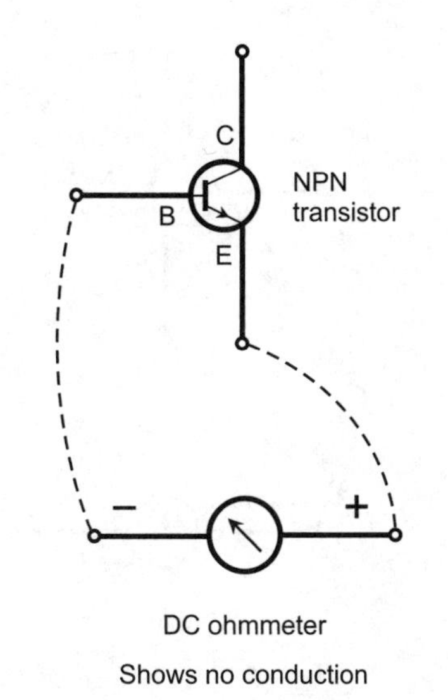

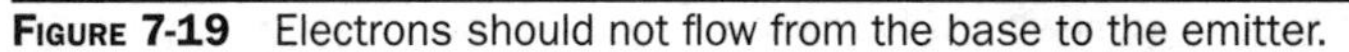

FIGURE 7-19 Electrons should not flow from the base to the emitter.

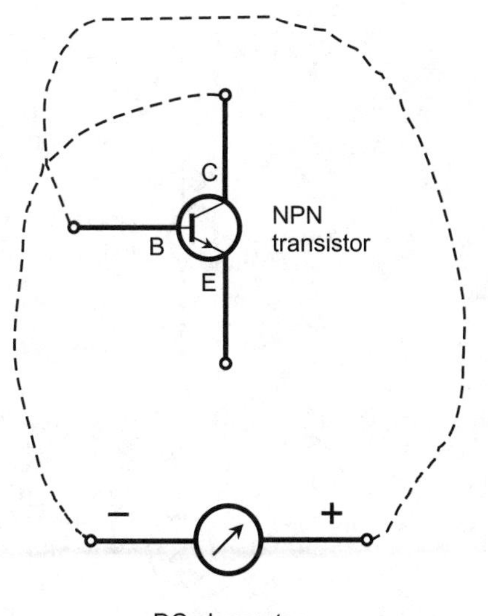

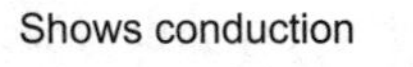

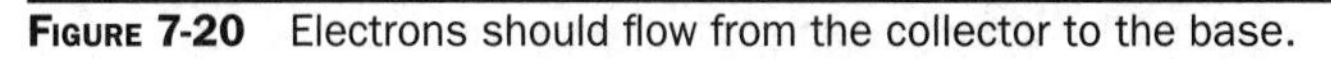

FIGURE 7-20 Electrons should flow from the collector to the base.

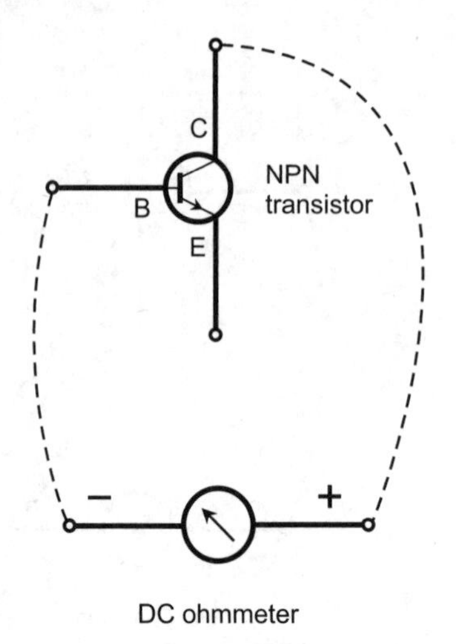

Shows no conduction

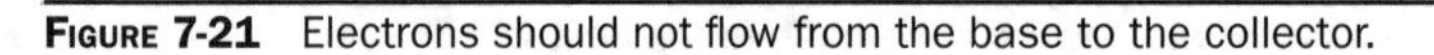

Figure 7-21 Electrons should not flow from the base to the collector.

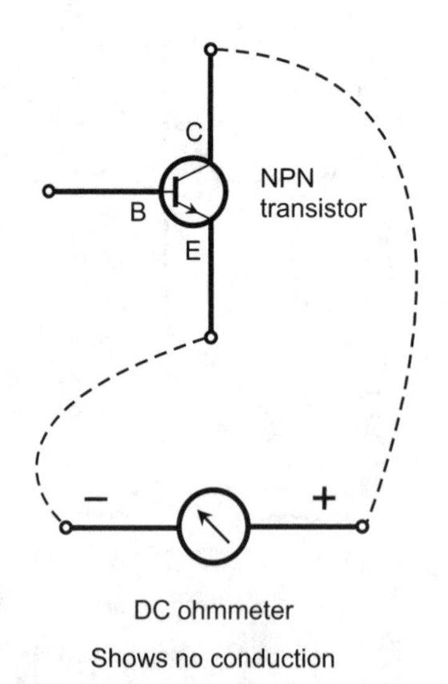

Shows no conduction

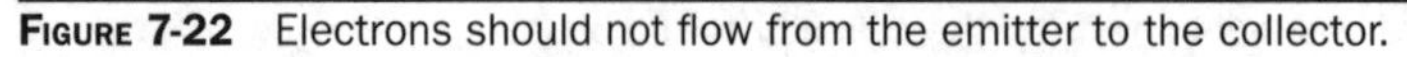

Figure 7-22 Electrons should not flow from the emitter to the collector.

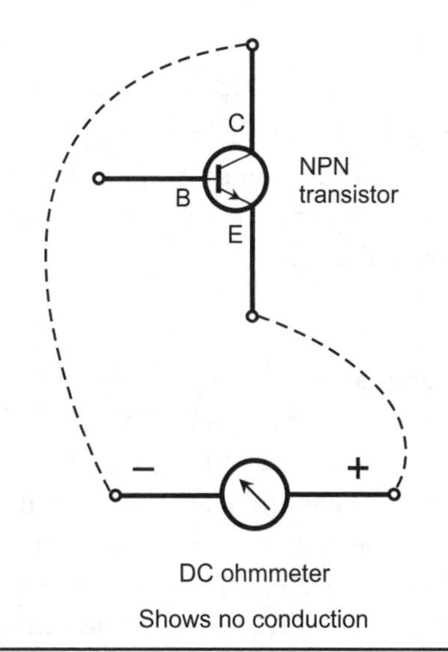

FIGURE 7-23 Electrons should not flow from the collector to the emitter.

Experiment 2: Check a JFET

You'll find JFETs at most Radio Shack retail outlets, and also on their website (www.radioshack.com), just as you can do with bipolar transistors. At the time of this writing, most JFET devices were available only through the website.

Get the Component

I used a P-channel JFET called NTE489, with Radio Shack catalog number 55052761. It's a tiny device, like the 2N222, so you might need strong glasses and, for sure, a steady hand. If you can't find the exact component, anything similar will work fine. Make sure it's a P-channel device, not an N-channel device.

> **Heads Up!**
>
> This test will work with JFETs, but not with MOSFETs. All JFETs have P-N junctions between the gate and the channel. However, MOSFETs do not have P-N junctions at all. You'll see "infinity" ohms in both directions between the gate and the channel in a MOSFET because the gate is insulated from the channel.

Prepare for Testing

Before you give your ohmmeter another workout, make sure that you know the locations of the source, gate, and drain on the JFET case. Radio Shack's JFETs come in packages with the lead placement guide printed on the back. Once you know where the leads are, keep the component "in isolation" so that you'll get true results when you check for conduction.

Do the Deed!

As you did with the bipolar transistor, hold the meter probe tips up against the JFET leads in all six combinations. Figures 7-24 through 7-29 show the connections and indicate the results you should see with a good P-channel JFET. Moderate resistance indicates conduction, just as with any other P-N junction. Don't worry about the exact ohmic value; your meter generates current that creates an "apparent resistance" in the junction. "Infinite" resistance, in which the meter needle does not budge (if you use an analog meter, as I did), however, definitely indicates non-conduction.

1. Electrons should flow from the gate to the source (Fig. 7-24).
2. Electrons should not flow from the source to the gate (Fig. 7-25).
3. Electrons should not flow from the drain to the gate (Fig. 7-26).
4. Electrons should flow from the gate to the drain (Fig. 7-27).
5. Electrons should flow from the source to the drain (Fig. 7-28).
6. Electrons should flow from the drain to the source (Fig. 7-29).

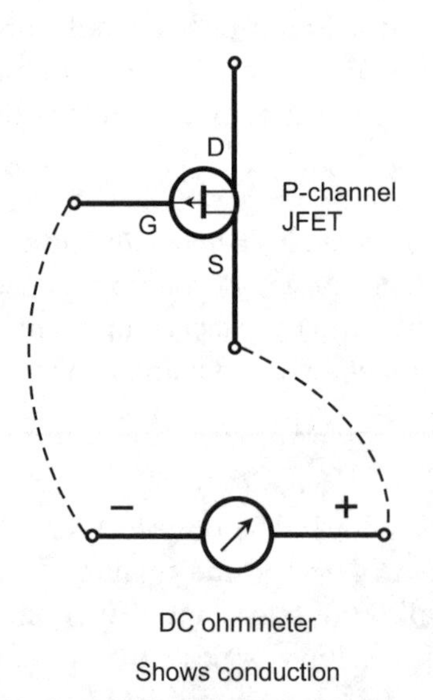

FIGURE 7-24 Electrons should flow from the gate to the source.

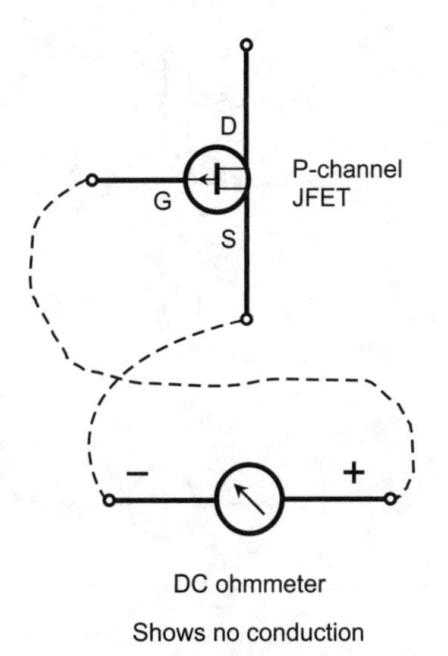

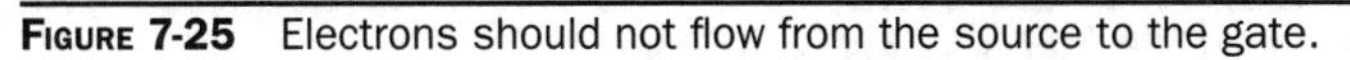

FIGURE 7-25 Electrons should not flow from the source to the gate.

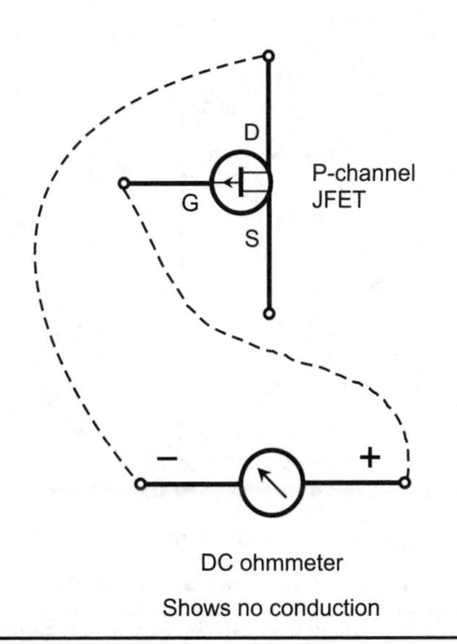

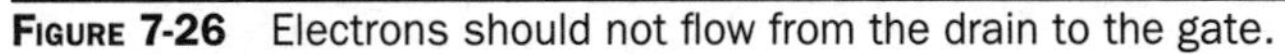

FIGURE 7-26 Electrons should not flow from the drain to the gate.

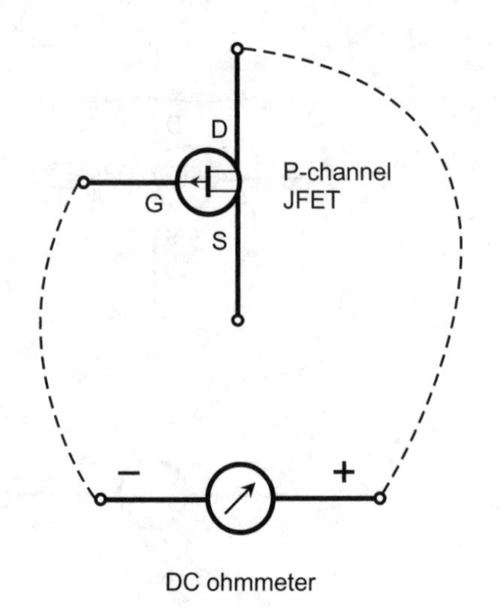

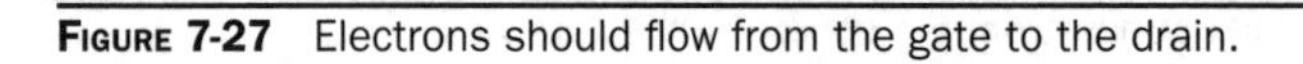

FIGURE 7-27 Electrons should flow from the gate to the drain.

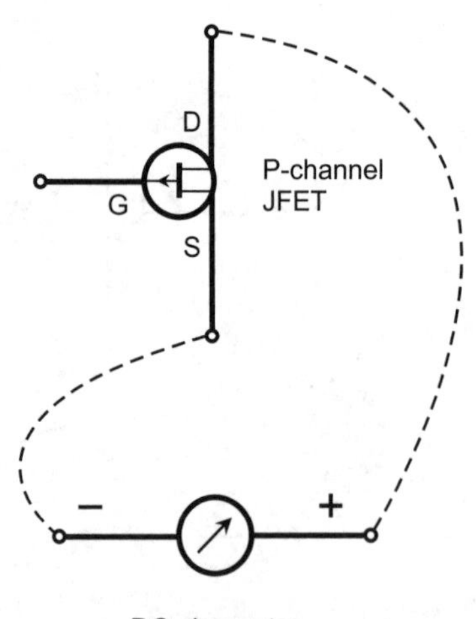

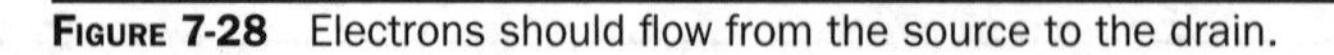

FIGURE 7-28 Electrons should flow from the source to the drain.

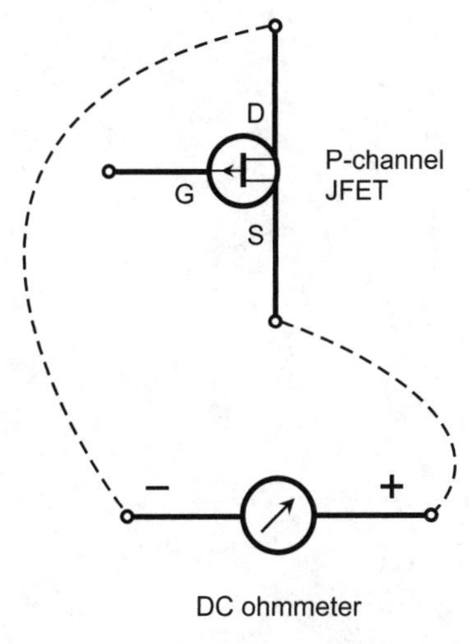

FIGURE 7-29 Electrons should flow from the drain to the source.

CHAPTER 8

Integrated Circuits and Digital Basics

Most *integrated circuits* (ICs), also known as *chips*, are tiny boxes with protruding metal terminals called *pins*. In schematic diagrams, engineers represent ICs as triangles or rectangles. *Digital logic*, also called simply *logic*, is the "reasoning" used by electronic machines. Engineers also use the term *logic* in reference to the circuits that make up digital devices and systems.

Advantages of IC Technology

Integrated circuits have advantages over discrete components (individual transistors, diodes, capacitors, and resistors). The most important considerations follow.

Compactness

An IC is far more compact than an equivalent circuit made from discrete components. Integrated circuits allow for the construction of more sophisticated systems in smaller packages than discrete components do.

High Speed

The interconnections between internal IC components are tiny, making high switching speeds possible. As you reduce the time it takes for charge carriers to get from one component to another, you increase the number of operations that a system can do within a given span of time, and you reduce the time it takes for the system to perform complicated operations.

Low Power Consumption

Integrated circuits consume less power than equivalent discrete-component circuits. This advantage makes a huge difference in battery-operated systems. In addition, because ICs use minimal current, they produce less heat than their discrete-component equivalents, resulting in improved efficiency.

Reliability

Integrated circuits fail less often, per component-hour of use, than systems built up from discrete components. The lower failure rate results from the fact that all component interconnections are sealed within the IC case, preventing the intrusion of dust, moisture, or corrosive gases. Therefore, IC-based systems suffer less downtime than discrete-component systems do.

Ease of Maintenance

Integrated-circuit technology minimizes hardware maintenance costs and streamlines maintenance procedures. Many appliances use sockets for ICs, and replacement involves nothing more than finding the faulty IC, unplugging it, and plugging in a new one.

Modular Construction

Modern IC appliances use *modular construction*, in which individual ICs perform defined functions within a circuit board. The circuit board or *card* fits into a socket and has a specific purpose. Repair technicians, using computers programmed with customized software, locate the faulty card, remove it, and replace it with a new one.

Limitations of IC Technology

No technological advancement comes without a downside. Integrated circuits have limitations that engineers must consider when designing an electronic device or system.

Inductors Impractical

While some components are easy to fabricate onto chips, others, notably inductors (coils), defy the IC manufacturing process. Devices using ICs must be designed to work with inductors external to the ICs themselves. However, resistance-capacitance (*RC*) circuits can do most things that inductance-capacitance (*LC*) circuits can do, and *RC* circuits are easy to etch onto IC chips.

High Power Impossible

In general, manufacturers can't fabricate a high-power amplifier onto a chip. High-power operation necessitates a certain minimum physical mass and volume because the components generate a lot of heat and that bulk and mass help to conduct and radiate the heat away.

Linear ICs

A *linear IC* processes *analog signals*, such as voices and music. The term *linear* arises from the fact that the amplification factor remains constant as the input amplitude

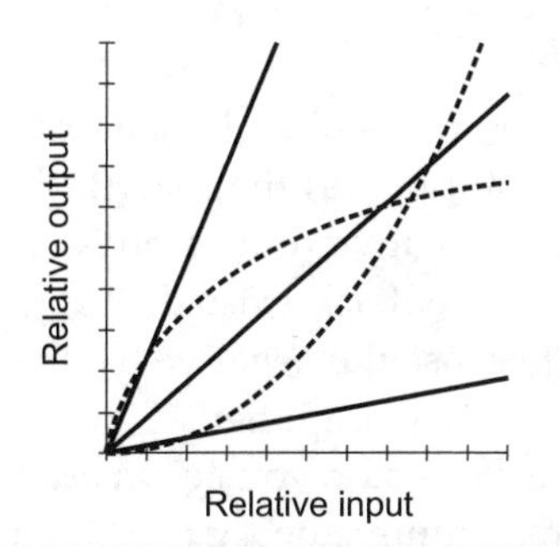

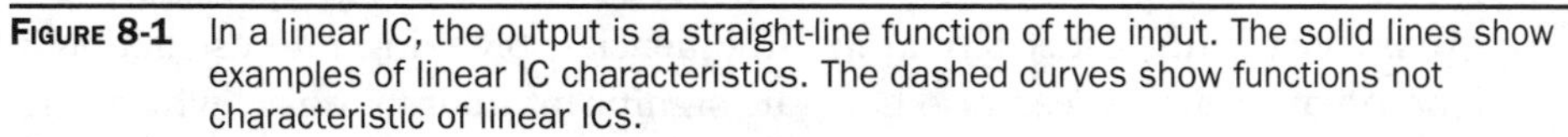

FIGURE 8-1 In a linear IC, the output is a straight-line function of the input. The solid lines show examples of linear IC characteristics. The dashed curves show functions not characteristic of linear ICs.

varies. In technical terms, the output signal strength constitutes a *linear function* of the input signal strength, as shown by any of the three solid, straight lines in the graph of Fig. 8-1. The dashed lines show *nonlinear functions*; they are curves.

Operational Amplifier

An *operational amplifier* (or *op amp*) comprises transistors, resistors, diodes, and capacitors interconnected to produce high gain over a wide range of frequencies. Some ICs contain two or more op amps, so you'll hear about *dual op amps* or *quad op amps*. Some ICs contain one or more op amps in addition to other circuits.

An op amp has two inputs, one *noninverting* and one *inverting*, and one output. When a signal goes into the noninverting input, the output wave emerges in phase coincidence with the input wave. When a signal goes into the inverting input, the output wave appears in phase opposition with respect to the input wave.

> **Tip**
>
> An op amp has two power-supply connections, one for the emitters of the internal transistors (V_{ee}) and one for the collectors (V_{cc}). The schematic symbol is a triangle (Fig. 8-2). The power supply produces well-defined negative and positive voltages relative to ground potential.

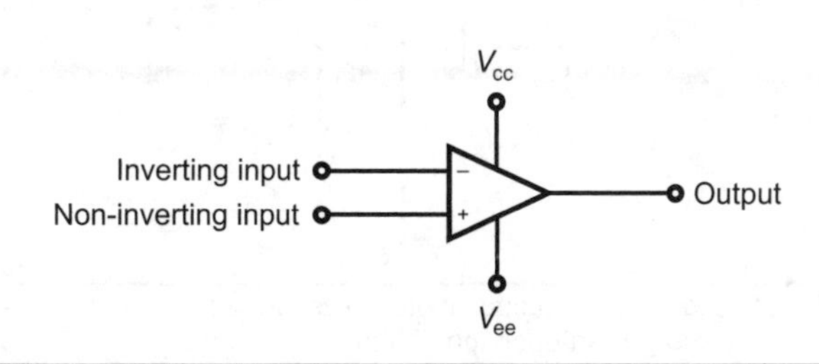

FIGURE 8-2 Schematic symbol for an op amp.

Op Amp Feedback and Gain

One or more external resistors determine the gain of an op amp. Normally, you place a resistor between the output and the inverting input to obtain a *closed-loop configuration*. The feedback is negative (out of phase), causing the gain to remain lower than it would be if no feedback existed. As you reduce the value of this resistor, the gain decreases because the negative feedback increases. Figure 8-3 is a schematic diagram of a closed-loop amplifier.

If you remove the feedback resistor, you get an *open-loop configuration*, in which the op amp produces its maximum rated gain. Open-loop op amps sometimes exhibit instability, especially at low frequencies, breaking into oscillation. Open-loop op-amp circuits can also generate significant internal noise, which can cause trouble in some applications.

If you install an *RC* combination in the feedback loop of an op amp, the gain depends on the input-signal frequency. Using specific values of resistance and capacitance, you can make a frequency-sensitive filter that provides any of four different characteristics, as shown in Fig. 8-4:

1. A *lowpass response* that favors low frequencies (at A).
2. A *highpass response* that favors high frequencies (at B).
3. A *peak* that produces maximum gain at a single frequency (at C).
4. A *notch* that produces minimum gain at a single frequency (at D).

Op Amp Differentiator

A *differentiator* is a circuit whose instantaneous output amplitude varies in direct proportion to the rate at which the input amplitude changes. In the mathematical

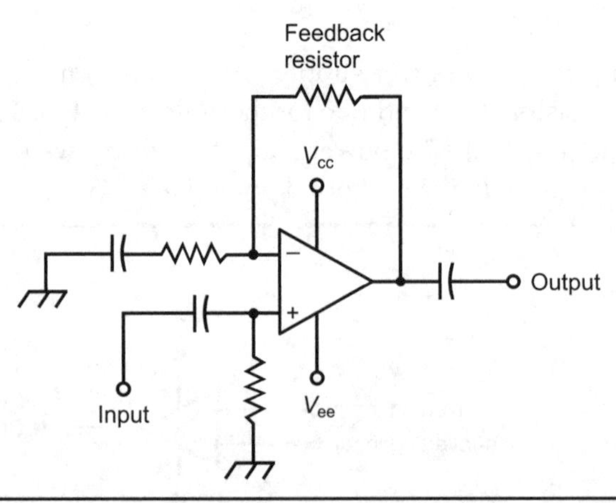

FIGURE 8-3 A closed-loop op amp circuit with negative feedback. If you remove the feedback resistor, you get an open-loop circuit.

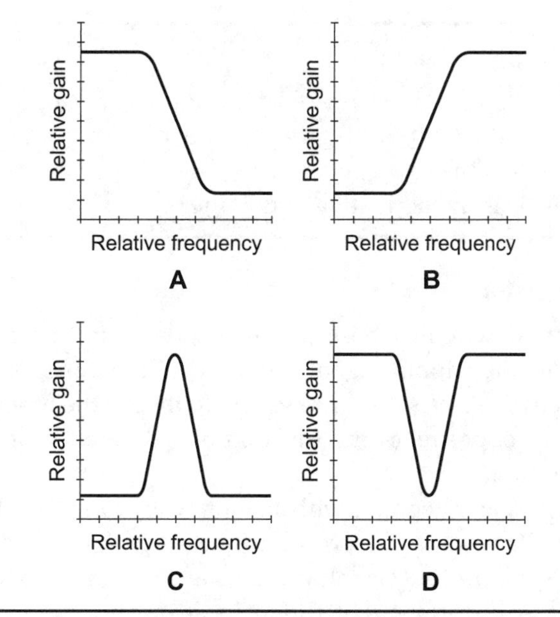

FIGURE 8-4 Gain-versus-frequency response curves. At A, lowpass; at B, highpass; at C, peak; at D, notch.

sense, the circuit *differentiates* the input signal wave function. Op amps lend themselves to use as differentiators. Figure 8-5 shows an example.

When the input to a differentiator is a constant DC voltage, the output equals zero (no signal). When the instantaneous input amplitude increases, the output is a positive DC voltage. When the instantaneous input amplitude decreases, the output is a negative DC voltage. If the input amplitude fluctuates periodically, the output voltage varies according to the *instantaneous rate of change* (mathematical *derivative*) of the instantaneous input-signal amplitude. You'll observe an output signal with the same frequency as that of the input signal, although the waveform might differ.

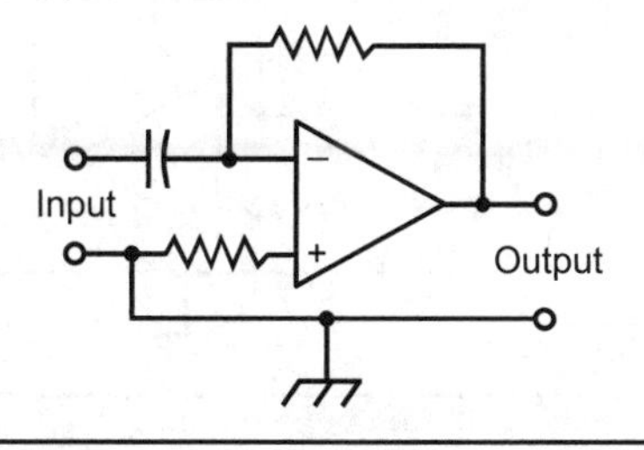

FIGURE 8-5 A differentiator circuit that uses an op amp.

Here's a Twist!

In a differentiator, a pure sinusoidal input produces a pure sinusoidal output, but the phase shifts 90° to the left (1/4 cycle earlier, or leading, in time). This wave represents the *cosine function*. Do you recall from your calculus courses that the cosine is the derivative of the sine?

Op Amp Integrator

An *integrator* is a circuit whose instantaneous output amplitude is proportional to the accumulated input signal amplitude, as a function of time. The circuit mathematically *integrates* the input signal. In theory, the function of an integrator is the inverse, or opposite, of the function of a differentiator. Figure 8-6 shows an op amp integrator.

If you supply an integrator with an input waveform that fluctuates periodically, the output voltage varies according to the *integral*, or *antiderivative*, of the input voltage. You get an output signal with the same frequency as that of the input signal, although the waveform might differ. In an integrator, a pure sine-wave input produces a pure sine-wave output, but the phase is shifted 90° to the right (1/4 cycle later, or lagging, in time).

Voltage Regulator

A *voltage regulator IC* acts to control the output voltage of a power supply. This feature is important with precision electronic equipment. You can find voltage-regulator ICs with a vast variety of voltage and current ratings for diverse applications.

Heads Up!

Typical voltage regulator ICs have three terminals, just as most transistors do. Because of this resemblance, these ICs are sometimes mistaken for power transistors. Of course, if you install a power transistor in place of a voltage regulator or vice versa, you should not expect a good outcome!

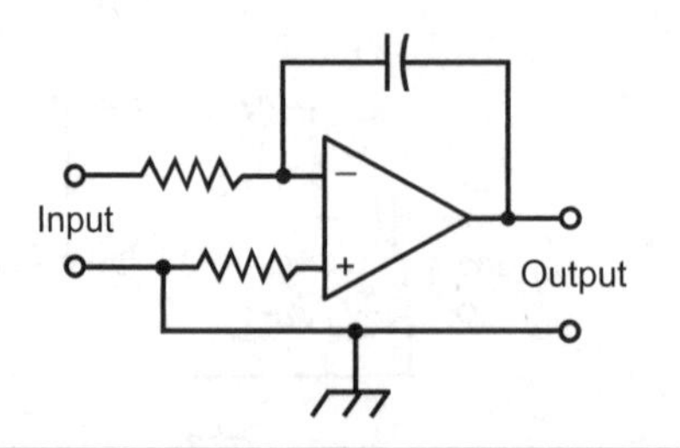

FIGURE 8-6 An integrator circuit that uses an op amp.

Timer

A *timer IC* is a specialized oscillator that produces a delayed output. You can tailor the delay time to suit a particular device. The delay is generated by counting the number of oscillator pulses; the length of the delay can be adjusted by means of external resistors and capacitors. Timer ICs are used in digital frequency counters, in which you need to provide a precise, constant, and predictable time "window."

Multiplexer

A *multiplexer IC* allows you to combine several different signals in one channel by means of a process called (you guessed it!) *multiplexing*. An analog multiplexer can also be used in reverse; then it works as a *demultiplexer*, separating the content of a multiplexed channel into its individual signals. You'll sometimes hear or read about a *multiplexer/demultiplexer IC*, which has both types of circuit on a single chip.

Comparator

A *comparator IC* has two inputs. It compares the voltages that appear at those inputs, which you call input A and input B. If the voltage at A significantly exceeds the voltage at B, the output equals about +5 V, giving you the logic 1, or high state. If the voltage at A is less than or equal to the voltage at B, you get an output voltage of about +2 V or less, yielding the logic 0, or low state.

Some comparators can switch between low and high states at a rapid rate, while others are slow. Some have low input impedance, and others exhibit high input impedance. Some are intended for AF or low-frequency RF use; others work well in high-frequency RF applications. Voltage comparators can actuate, or *trigger*, other devices, such as relays, alarms, and electronic switching circuits.

Digital ICs

The most common type of *digital IC*, also called a *digital-logic IC*, operates using two discrete states: high (logic 1) and low (logic 0). Digital ICs contain massive arrays of logic gates that perform switching operations at high speed.

Transistor-Transistor Logic

In *transistor-transistor logic* (TTL), arrays of bipolar transistors, some with multiple emitters, operate on DC pulses. Figure 8-7 illustrates the internal details of a basic TTL gate using two NPN bipolar transistors, one of which is a *dual-emitter* device. In TTL, the transistors are always either cut off or saturated. No analog amplification can occur; analog input signals are "swamped" by the digital states. Because of this well-defined duality, TTL systems offer excellent immunity to external analog noise.

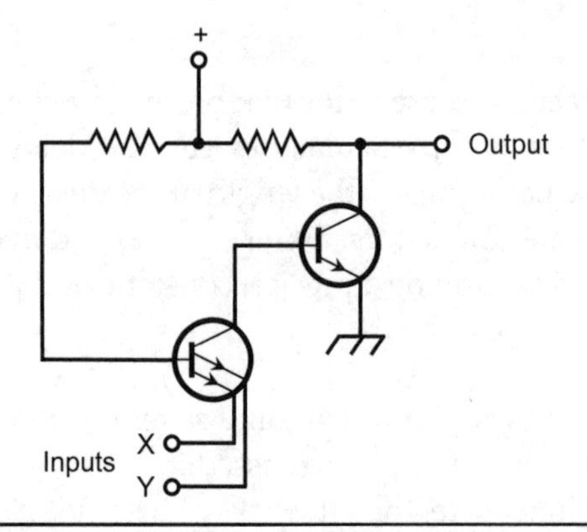

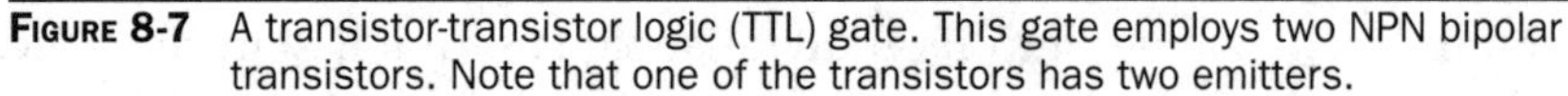

Figure 8-7 A transistor-transistor logic (TTL) gate. This gate employs two NPN bipolar transistors. Note that one of the transistors has two emitters.

Emitter-Coupled Logic

Emitter-coupled logic (ECL) is another common bipolar-transistor scheme. In an ECL device, the transistors don't operate at saturation, as they do with TTL. This biasing scheme increases the speed of ECL relative to TTL, but it also increases the vulnerability to analog noise. That's because, when transistors don't operate in saturation, they respond to analog disturbances as well as the desired digital signals. Figure 8-8 shows the internal details of a basic ECL gate using four NPN bipolar transistors.

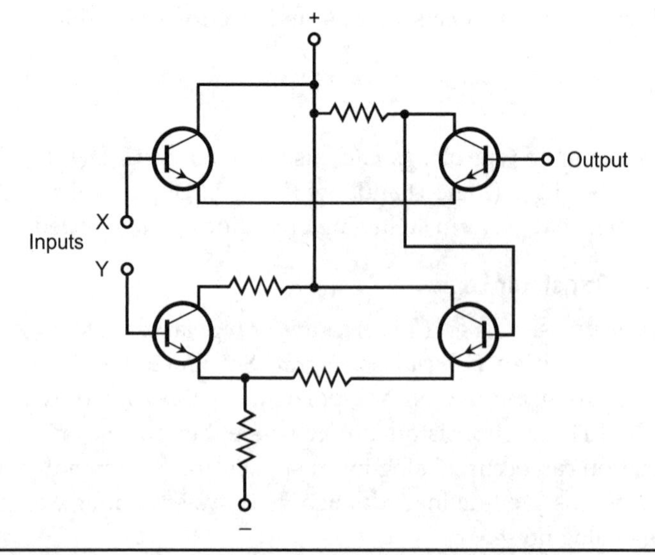

Figure 8-8 An emitter-coupled logic (ECL) gate using four NPN bipolar transistors.

Metal-Oxide-Semiconductor Logic

Digital ICs can be constructed using metal-oxide-semiconductor (MOS) technology. *N-channel MOS* (NMOS) *logic*, pronounced "EN-moss logic," offers simplicity of design, along with high operating speed. *P-channel MOS logic*, pronounced "PEA-moss logic," resembles NMOS logic, but generally runs a bit slower. You can summarize the construction details of these two schemes as follows:

1. An NMOS IC is the counterpart of a discrete-component circuit that uses N-channel MOSFETs.
2. A PMOS IC is the counterpart of a discrete-component circuit that uses P-channel MOSFETs.

Complementary-metal-oxide-semiconductor (CMOS) *logic*, pronounced "SEA-moss logic," employs both N-type and P-type silicon on a single chip, analogous to using both N-channel and P-channel MOSFETs in a discrete-component circuit. The advantages of CMOS technology include minimal current drain, high operating speed, and immunity to analog noise.

All forms of MOS ICs require careful handling to prevent destruction by electrostatic discharges. The precautions are the same as those for handling discrete MOSFETs. All personnel who work with MOS ICs should ground their bodies by wearing metal wrist straps connected to electrical ground. It also helps to make sure that the humidity in the lab does not get too low.

Tip

When you store MOS ICs, you can push the pins into conductive foam specifically manufactured for that purpose. This precaution helps to protect the MOS material from electrostatic charge buildup in storage.

Component Density

The number of elements per chip in an IC gives you a figure for *component density*. The past several decades have seen a steady increase in the number of components that manufacturers can fabricate on a single chip. That trend will likely continue for some time, until the structure of matter itself forces developers to refine and deploy new technologies and methods of manufacture.

Small-Scale Integration

Small-scale integration (SSI) has fewer than 10 transistors on a chip. These devices can carry the largest currents of any IC type because low component density translates into relatively large volume and mass per component. Small-scale integration finds application in voltage regulators and other moderate-power systems.

Medium-Scale Integration

In *medium-scale integration* (MSI), you'll find 10 to 100 transistors per chip. This density allows for considerable miniaturization, but does not constitute a high level of component density, relatively speaking, these days. An advantage of MSI is the fact that individual logic gates can carry fairly large currents in some applications. Both bipolar and MOS technologies can be adapted to MSI.

Large-Scale Integration

In *large-scale integration* (LSI), you have 100 to 1000 transistors per semiconductor chip, a full *order of magnitude* (a factor of 10 times or power of 10) more dense than MSI. Electronic wristwatches, single-chip calculators, microcontrollers, and microcomputers use LSI chips.

Very-Large-Scale Integration

Very-large-scale integration (VLSI) devices have from 1000 to 1,000,000 transistors per chip, up to three orders of magnitude more dense than LSI. High-end microcomputers, microcontrollers, and memory chips are made using VLSI.

Ultra-Large-Scale Integration

You'll sometimes hear of *ultra-large-scale integration* (ULSI). Devices of this kind have more than 1,000,000 transistors per chip. The principal uses for ULSI technology include high-level computing, supercomputing, military and medical robotics, and artificial intelligence (AI).

IC Memory

Binary data in the form of high and low states (1 and 0) can be stored in memory chips that take a wide variety of physical forms. Some IC memory chips require a continuous source of backup voltage or they'll lose their data. Others can hold the data indefinitely in the absence of backup voltage. In electronic devices, you encounter two other major types of memory: *random-access* and *read-only*.

Random-Access Memory

A *random-access memory* (RAM) chip stores binary data in *arrays*. The data can be *addressed* (selected) from anywhere in the matrix. Data is easily retrieved, changed, and stored back. Engineers sometimes call a RAM chip *read/write memory*. There are two major categories of RAM: *dynamic RAM* (DRAM) and *static RAM* (SRAM).

1. A DRAM chip contains transistors and capacitors, and data is stored as charges on the capacitors. The charge must be replenished frequently, or it will vanish.

2. An SRAM chip uses a specialized type of logic gate called a *flip-flop* to store the data. This gets rid of the need for constant replenishing of the charge, but SRAM ICs require more elements than DRAM chips to store a given amount of data.

With any RAM chip, the data will be lost when you remove power unless you provide a means of *memory backup*. The most common memory-backup scheme involves the use of a small lithium battery with a long shelf life.

> **Wow!**
>
> Today's memory chips need so little current to store data that a backup battery lasts as long in the circuit as it would sitting on a shelf.

Read-Only Memory

The data in a *read-only memory* (ROM) chip can be easily accessed but not written over. A standard ROM chip is programmed at the factory with *firmware*. Some ROM chips allow you to program and reprogram them yourself.

If the data in memory disappears when you remove all sources of power, you have *volatile memory*. If the data is retained indefinitely after you remove the power, then you have *nonvolatile memory*. While the data in most RAM chips is volatile, the data in ROM chips is nonvolatile.

An *erasable programmable read-only memory* (EPROM) chip is a ROM device that you can reprogram by exposure to ultraviolet (UV). You can recognize an EPROM chip by the presence of a transparent window through which UV rays are focused to erase the data.

> **Tip**
>
> The data in some EPROM chips can be erased by electrical means. Such an IC is called an *electrically erasable programmable read-only memory* (EEPROM) chip.

Microcomputers and Microcontrollers

A *microcomputer* is a small computer manufactured in a single IC package or on a printed-circuit board with other components. The Raspberry Pi is a common example. A *microcontroller* is a specialized microcomputer, designed to regulate the operation of electrical and electromechanical devices such as robots. Examples include the *PIC* and the *Arduino*.

Microcomputers vary in processing power and memory capacity, depending on the intended use. Simple microcomputers are available at the retail level for a few dollars. Some microcomputers have LCDs and small keypads that allow encoded data entry. Smartphones that include Internet access, video cameras, and gaming applications contain microcomputer chips. Some microcontrollers can accept voice instructions.

Microcontrollers allow machines to perform complex, repetitive tasks. Low-end and intermediate microcontrollers are used in automobiles and home appliances. For example, a microcontroller can be programmed to switch on an oven, heat the food to a prescribed temperature for a certain length of time, and then switch the oven off. Microcontrollers can regulate the operation of an automobile engine to enhance efficiency and gasoline mileage.

High-end microcontrollers can perform some critical operations in motor vehicles, boats, and airplanes. For example, an automobile might sense a nearby vehicle in the "blind spot" on a freeway, and prevent the driver from changing lanes in such a way as to cause an accident.

Did You Know?

Medical electronics offers one of the most exciting applications of high-end microcontrollers. A microcontroller can produce regulated electrical impulses to govern the behavior of erratically functioning body organs, and even move the muscles of people who would otherwise lack mobility.

Boolean Algebra

Boolean algebra is a system of digital logic using the numbers 0 and 1 with the operations AND (multiplication), OR (addition), and NOT (negation). Combinations of these operations give you two more, called NAND (NOT AND) and NOR (NOT OR). This system, which gets its name from the nineteenth-century British mathematician *George Boole*, plays a vital role in the design of digital electronic circuits.

- The AND operation, also called *logical conjunction*, operates on two or more quantities. It's commonly denoted as an asterisk, for example X * Y.
- The NOT operation, also called *logical inversion* or *logical negation*, operates on a single quantity. It's commonly denoted as a minus sign (–), for example –X.
- The OR operation, also called *logical disjunction*, operates on two or more quantities. It's commonly denoted as a plus sign (+), for example X + Y.

Table 8-1 breaks down all the possible input and output values for the above-described Boolean operations, in which 0 indicates "falsity" and 1 indicates "truth."

TABLE 8-1 The Basic Boolean Operations NOT, AND, and OR

X	Y	–X	X * Y	X + Y
0	0	1	0	0
0	1	1	0	1
1	0	0	0	1
1	1	0	1	1

Theorems

Table 8-2 shows some logic equations that hold true for all possible values of the *logical variables* X, Y, and Z. Such facts are called *theorems*. Statements on either side of the equals (=) sign are *logically equivalent*, meaning that one is true *if and only if (iff)* the other is true. For example, the statement X = Y means "If X then Y, and if Y then X." Boolean theorems can allow you to simplify complicated *logic functions*,

TABLE 8-2 Some Common Theorems in Boolean Algebra

Logical Expression	Technical Term
X + 0 = X	OR identity
X * 1 = X	AND identity
X + 1 = 1	
X * 0 = 0	
X + X = X	
X * X = X	
–(–X) = X	Double negation
X + (–X) = 1	
X * (–X) = 0	Contradiction
X + Y = Y + X	Commutative property of OR
X * Y = Y * X	Commutative property of AND
X + (X * Y) = X	
X * (–Y) +Y = X + Y	
(X + Y) + Z = X + (Y + Z)	Associative property of OR
(X * Y) * Z = X * (Y * Z)	Associative property of AND
X * (Y + Z) = (X * Y) + (X * Z)	Distributive property
–(X +Y) = (–X) * (–Y)	DeMorgan's Theorem
–(X * Y) = (–X) + (–Y)	DeMorgan's Theorem

facilitating the construction of circuits that perform specific digital operations using the minimum number of switches.

Positive versus Negative Logic

In *positive logic*, a circuit with a voltage of +3 V to + 6 V DC represents the binary digit 1 (called the *high state*), while the binary digit 0 appears as little or no DC voltage (called the *low state*).

Some circuits employ *negative logic*, in which little or no DC voltage (low state) represents logic 1, while the positive DC voltage (high state) represents logic 0. In another form of negative logic, the digit 1 appears as −3 V to −6 V DC (the low state) and the digit 0 appears as little or no DC voltage (the high state). As a hobbyist, you'll rarely have to concern yourself with negative logic.

Logic Gates

All digital devices employ switches that perform logical operations. These switches, called *logic gates*, have one or more inputs and (usually) a single output.

- A *logical inverter*, also called a *NOT gate*, has one input and one output. It reverses, or inverts, the state of the input. If the input equals 1, then the output equals 0. If the input equals 0, then the output equals 1.
- An *OR gate* can have two or more inputs (although it usually has only two). If both, or all, of the inputs equal 0, then the output equals 0. If any of the inputs equal 1, then the output equals 1.
- An *AND gate* can have two or more inputs (although it usually has only two). If both, or all, of the inputs equal 1, then the output equals 1. If any of the inputs equal 0, then the output equals 0.
- An OR gate can be followed by a NOT gate to form a *NOT-OR gate*, more often called a *NOR gate*. If both, or all, of the inputs equal 0, then the output equals 1. If any of the inputs equal 1, then the output equals 0.
- An AND gate can be followed by a NOT gate to form a *NOT-AND gate*, more often called a *NAND gate*. If both, or all, of the inputs equal 1, then the output equals 0. If any of the inputs equals 0, then the output equals 1.
- An *exclusive OR gate*, also called an *XOR gate*, has two inputs and one output. If the two inputs have the same state (both 1 or both 0), then the output equals 0. If the two inputs have different states, then the output equals 1.

Table 8-3 summarizes the functions of the above-defined logic gates. Figure 8-9 illustrates the schematic symbols that you should use to represent these gates in circuit diagrams, assuming a single input for the NOT gate and two inputs for the others.

TABLE 8-3 Logic Gates and Their Characteristics

Gate Type	Number of Inputs	Remarks
NOT	1	Changes state of input.
OR	2 or more	Output high if any inputs are high. Output low if all inputs are low.
AND	2 or more	Output low if any inputs are low. Output high if all inputs are high.
NOR	2 or more	Output low if any inputs are high. Output high if all inputs are low.
NAND	2 or more	Output high if any inputs are low. Output low if all inputs are high.
XOR	2	Output high if inputs differ. Output low if inputs are the same.

Clocks

In electronics, the term *clock* refers to a circuit that generates pulses at high speed and precise, constant time intervals. A clock sets the tempo for the operation of digital devices. In a computer, a clock acts as a metronome for the microprocessor. Engineers express clock speeds in *hertz* (Hz), where 1 Hz equals one pulse per second. Higher-frequency units are expressed just as they are with analog signals:

- A *kilohertz* (kHz) equals 1000 or 10^3 pulses per second
- A *megahertz* (MHz) equals 1,000,000 or 10^6 pulses per second
- A *gigahertz* (GHz) equals 1,000,000,000 or 10^9 pulses per second
- A *terahertz* (THz) equals 1,000,000,000,000 or 10^{12} pulses per second

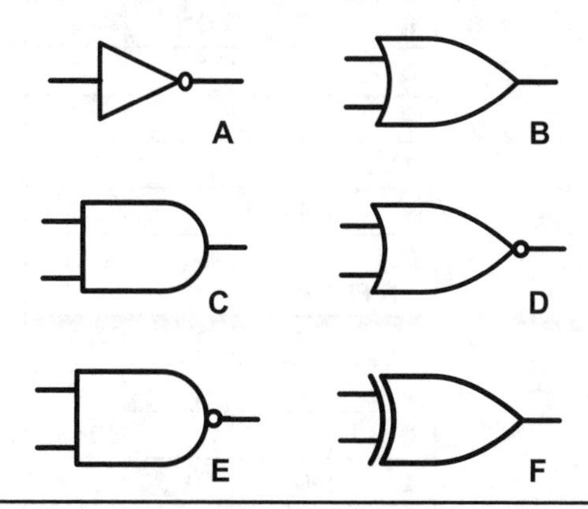

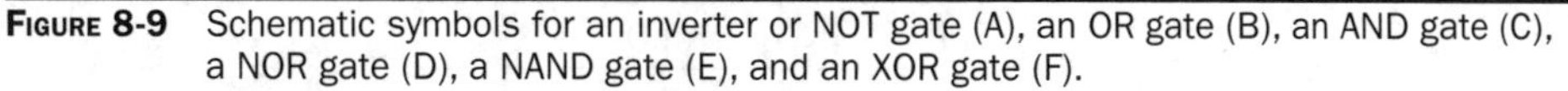

FIGURE 8-9 Schematic symbols for an inverter or NOT gate (A), an OR gate (B), an AND gate (C), a NOR gate (D), a NAND gate (E), and an XOR gate (F).

Flip-Flops

A *flip-flop* is a circuit constructed from logic gates, known collectively as a *sequential gate*. The term *sequential* comes from the fact that the output depends not only on the states of the circuit at any given instant in time, but also on the state or states that existed just before it.

A flip-flop has two states, called *set* and *reset*. Usually, the set state corresponds to logic 1 (high), and the reset state corresponds to logic 0 (low). In schematic diagrams, a flip-flop appears as a rectangle with two or more inputs and two outputs and the letters FF at the top. Several different types of flip-flops exist, as follows:

- In an *R-S flip-flop*, the inputs are labeled R (reset) and S (set). Engineers call the outputs Q and −Q. (Often, rather than −Q, you'll see Q′, or perhaps Q with a line over it.) Table 8-4A shows the input and output states. If R = 0 and S = 0, the output states stay at the values they've attained for the moment. If R = 0 and S = 1, then Q = 1 and −Q = 0. If R = 1 and S = 0, then Q = 0 and −Q = 1. When S = 1 and R = 1, the circuit becomes unpredictable.
- In a *synchronous flip-flop*, the states change when triggered by the signal from an external clock. In *static triggering*, the outputs change state only when the clock signal is either high or low. Some engineers call this circuit a *gated flip-flop*. In *positive-edge triggering*, the outputs change state at the instant the clock signal changes from low to high. In *negative-edge triggering*, the outputs change state at the instant the clock signal changes from high to low.

Table 8-4 Flip-Flop States

A: R-S Flip-Flop			
R	**S**	**Q**	**−Q**
0	0	Q	−Q
0	1	1	0
1	0	0	1
1	1	?	?
B: J-K Flip-Flop			
J	**K**	**Q**	**−Q**
0	0	Q	−Q
0	1	1	0
1	0	0	1
1	1	−Q	Q

- In a *master/slave* (M/S) *flip-flop*, the inputs are stored before the outputs can change state. This device comprises essentially two R-S flip-flops in series. You can call the first flip-flop the *master* and the second flip-flop the *slave*. The master functions when the clock output is high, and the slave acts during the next ensuing low portion of the clock output. The time delay prevents confusion between the input and output.
- The operation of a *J-K flip-flop* resembles the functioning of an R-S flip-flop, except that the J-K device has a predictable output when the inputs both equal 1. Table 8-4B shows the input and output states for this type of flip-flop. The output changes only when a triggering pulse is received.
- The operation of an *R-S-T flip-flop* resembles that of an R-S flip-flop, except that a high pulse at an additional T input causes the circuit to change state.
- A circuit called a *T flip-flop* has only one input. Each time a high pulse appears at the input, the output reverses its state. Note the difference between this type of circuit and a simple inverter (NOT gate)!

Counters

A *counter* tallies up digital pulses one by one. Each time the counter receives a high pulse, the binary number in its memory increases by 1. A *frequency counter* can measure the frequency of an AC wave or signal by tallying up the cycles over a certain fixed period of time such as 1 second. The circuit comprises a *gate* that begins and ends each counting cycle at defined intervals. (Don't confuse this type of gate with the logic gates described a few moments ago.) The counter's accuracy depends on the *gate time*, or how long the gate remains open to accept pulses for counting. As you increase the gate time in a frequency counter, the accuracy improves.

Experiment 1: Build an OR Gate

The most common type of OR gate has two inputs and one output. For any input or output terminal, two possible logic states exist: high (a few positive volts DC) and low (near 0 V DC). If both inputs are low, then the output is low. If either or both inputs are high, then the output is high.

Components

You can build a circuit to simulate the logical OR function with a couple of rectifier diodes, a 1/2-W resistor rated at 470 ohms, a 6-V lantern battery, a few clip leads, and your multimeter. You can get all of these components at Radio Shack. Figure 8-10 is the schematic diagram. The ground connection should go to the negative terminal of the battery (not shown here). The diode cathodes should go to the output and the non-grounded end of the resistor. The diode anodes serve as the gate inputs.

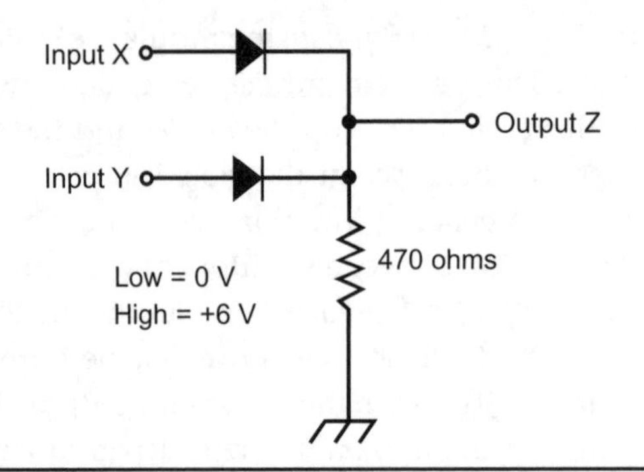

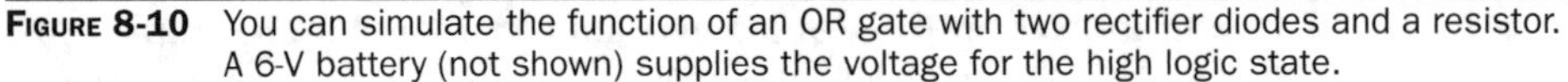

FIGURE 8-10 You can simulate the function of an OR gate with two rectifier diodes and a resistor. A 6-V battery (not shown) supplies the voltage for the high logic state.

> **Tip**
>
> If you like, you can "kludge" this circuit by crudely twisting the component leads together. You don't need a breadboard or perfboard unless you insist on having a physically attractive device.

How It Works

Depending on the voltages at the inputs, the diodes will either conduct or not conduct, producing a DC output voltage of either 0 V or +6 V. When you want to force an input low, short it to the negative battery terminal (ground) using a clip lead. When you want to force an input high, short it to the positive battery terminal. When you use your multimeter to check the voltages in all four possible cases, you should get the following results, which correspond to the data for the OR gate in Tables 8-1 and 8-3.

1. *Inputs X and Y both low*—The entire circuit is dead. No voltage exists anywhere; the battery might as well not exist! Output Z appears low along with every other point in the circuit. That checks out with the theoretical OR function: X = 0, Y = 0, and Z = 0.
2. *Input X low, input Y high*—In this case, electrons flow from ground through the resistor and diode Y to the positive battery terminal. Electrons also flow away from output Z through diode Y, creating a positive voltage at Z, so you see a logical high state there. That checks out with the theoretical OR function: X = 0, Y = 1, and Z = 1.
3. *Input X high, input Y low*—Electrons flow from ground through the resistor and diode X to the positive battery terminal. Electrons also flow away from output Z through diode X, creating a positive voltage at Z, so you see a logical high state there. It checks out with the theoretical OR function: X = 1, Y = 0, and Z = 1.

4. *Inputs X and Y both high*—Electrons flow from ground through the resistor and both diodes to the positive battery terminal. Electrons also flow away from output Z through both diodes, creating a positive voltage at Z, so you see a logical high state there. That checks out with the theoretical OR function: X = 1, Y = 1, and Z = 1.

> **Tip**
>
> Remember that in a rectifier diode, electrons can flow easily against the arrow (from cathode to anode), but not in the direction that the arrow points (from anode to cathode). This "one-way current gate" property makes the diodes act as switches, combining their effects to execute the logical OR function when connected as shown in Fig. 8-10.

Experiment 2: Build an AND Gate

The typical AND gate has two inputs and one output, and works with the same high and low logic states as the OR gate does. But the function itself radically differs! With an OR gate, three out of the four possible input combinations yield a high output. With an AND gate, only one out of the four possible input combinations produces a high output. When either or both inputs are low, the output is low. The output is high only when both inputs are high.

Components

You can simulate the AND function with the same components that you used to mimic the OR gate: two rectifier diodes, a 1/2-W resistor rated at 470 ohms, a 6-V lantern battery, some clip leads, and a multimeter. Figure 8-11 is a schematic diagram of a discrete-component AND gate circuit. At first glance, it looks like the circuit for

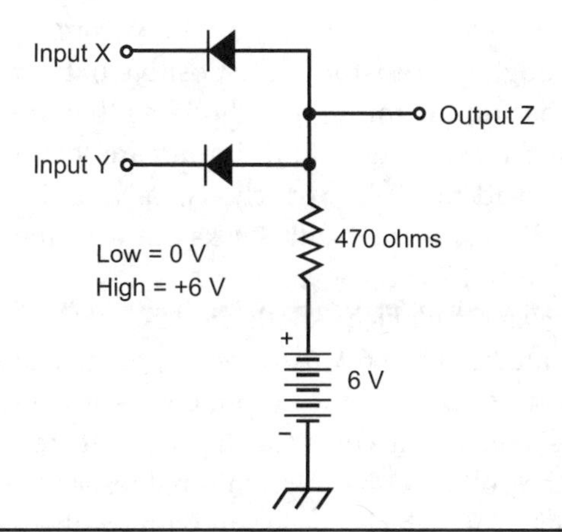

FIGURE 8-11 You can simulate the function of an AND gate with two rectifier diodes, a resistor, and a 6-V battery.

the OR gate, but critical differences exist! The diode polarities are reversed, and instead of going to ground, the non-diode end of the resistor is forced high by connecting it to the positive battery terminal. (As with the OR gate, the negative battery terminal serves as the ground connection.) The diode anodes go to the output and the non-battery end of the resistor. The diode cathodes are the inputs.

How It Works

You generate the high and low states in the same fashion with this circuit as you do with the OR circuit. When you want to force an input low, short it to the negative battery terminal (ground). When you want to force an input high, short it to the positive battery terminal. When you use your multimeter to check the voltages in all four possible input combination cases, you should get the following results, which correspond to the data for the AND gate in Tables 8-1 and 8-3.

1. *Inputs X and Y both low*—Electrons flow from both inputs through the diodes from cathode to anode, and onward through the resistor to the positive battery terminal. The non-battery end of the resistor is held to +0.6 V, the forward breakover voltage of silicon rectifier diodes. (The diodes have that amount of voltage drop when forward biased.) That's close enough to 0 V so that it qualifies as a low state, so the result checks out with the theoretical AND function: X = 0, Y = 0, and Z = 0.
2. *Input X low, input Y high*—Electrons flow from input X through its diode and onward through the resistor to the positive battery terminal. Input Y, held at +6 V, sends nothing through its diode because the anode is at +0.6 V. Diode Y does not conduct, so it acts as an open switch, and might as well not be there at all. Just as in the case in which both inputs are low, the non-battery end of the resistor is held to +0.6 V, which is logic low. The output is, therefore, low, and the result checks out with the theoretical AND function: X = 0, Y = 1, and Z = 0.
3. *Input X high, input Y low*—Electrons flow from input Y through its diode and along through the resistor to the positive battery terminal. Input X, held at +6 V, sends nothing through its diode, so that diode mimics an open switch. Just as in the case where both inputs are low, the non-battery end of the resistor is held to +0.6 V, which acts as a logical low state. The output is, therefore, low, and the result checks out with the theoretical AND function: X = 1, Y = 0, and Z = 0.
4. *Inputs X and Y both high*—Both diodes lack bias voltage; their anodes and cathodes are both at +6 V. As a result, neither diode conducts. Both diodes act as open switches, so the output comes straight from the non-battery end of the resistor. When you measure the voltage at Z, you should see +6 V because the voltmeter has high internal resistance (far more than 470 ohms). That's logic high, which checks out with the theoretical AND function: X = 1, Y = 1, and Z = 1.

CHAPTER 9

More Components and Techniques

As you gain experience in electronics, you'll want to know more about components, do more tests and projects, and buy more stuff. This chapter describes some of the most likely avenues into which you'll expand your interest and expertise.

Cells and Batteries

Engineers call a unit source of DC energy a *cell*. When you connect two or more cells in series, parallel, or series-parallel, you obtain a *battery*.

Zinc-Carbon Cells

Figure 9-1 is a "translucent" drawing of a *zinc-carbon cell*. The zinc forms the case, which serves as the negative electrode. A carbon rod constitutes the positive electrode. The electrolyte is a paste of manganese dioxide and carbon. Zinc-carbon cells don't cost much. They work well at moderate temperatures and in applications that demand moderate to high current.

Alkaline Cells

An *alkaline cell* has granular zinc as the negative electrode, potassium hydroxide as the electrolyte, and an element called a *polarizer* as the positive electrode. The construction resembles that of the zinc-carbon cell. Alkaline cells can work at lower temperatures than zinc-carbon cells can. Alkaline cells can deliver as much current as zinc-carbon cells can, but they last longer. They're the cell of choice for most consumer applications.

Transistor Batteries

A *transistor battery* consists of six tiny zinc-carbon or alkaline cells connected in series and enclosed in a box-shaped case about the size of a computer thumb drive. Each internal cell supplies 1.5 V, so the whole battery supplies 9 V. You can find

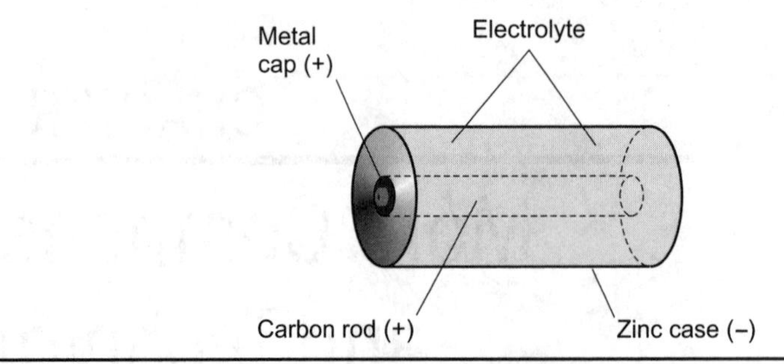

FIGURE 9-1 Internal structure of a zinc-carbon electrochemical cell.

transistor batteries in low-current electronic devices, such as wireless remote-control boxes and smoke detectors.

Lantern Batteries

A *lantern battery* has much greater mass than a dry cell or transistor battery, so it lasts far longer and can deliver a lot more current when the demand arises. Lantern batteries are usually rated at 6 V. They work well in small radios, such as low-power transceivers for medium-power needs. Most of these batteries comprise numerous size AA cells in a series-parallel arrangement.

> **Tip**
>
> Some serious electronics hobbyists dismantle lantern batteries to get at the individual AA cells inside. They work as well as the AA cells that you buy off store shelves, but they cost less per cell because of mass production for the battery. I recommend that you do this trick only if you have experience with cells and batteries; you can make a huge mess if you aren't careful!

Silver-Oxide Cells and Batteries

Silver-oxide cells are tiny, with a button-like shape. They supply 1.5 V, and offer excellent energy storage capacity per unit mass. They have a relatively *flat discharge curve* like the one in Fig. 9-2, which is considered ideal. Zinc-carbon and alkaline cells and batteries, in contrast, have *a declining discharge curve* (Fig. 9-3).

Lithium Cells and Batteries

Lithium cells gained popularity in the late 1970s and 1980s. You can find several variations in the chemical makeup of these cells. Most supply 1.5 V to 3.5 V, depending on the method of manufacture. Lithium cells and batteries can last for years in very-low-current applications such as the powering of a digital wristwatch. They provide high energy capacity per unit of mass.

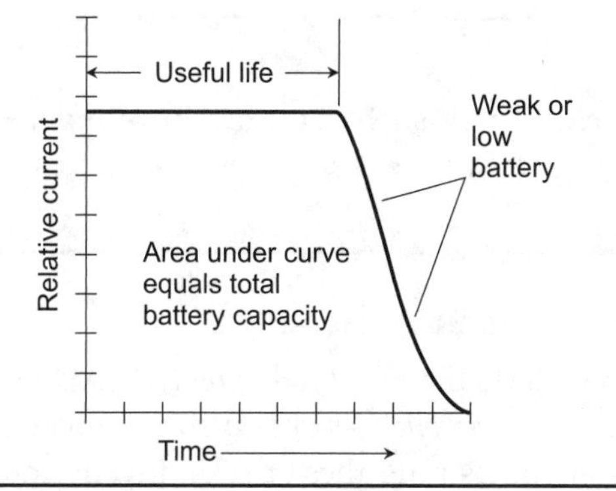

Figure 9-2 A flat discharge curve.

> **Did You Know?**
> One of the earliest uses for lithium batteries was to serve as memory backup for the microcomputers that control electronic devices, such as ham radios.

Lead-Acid Cells and Batteries

A typical *lead-acid cell* has a solution of sulfuric acid, along with a lead electrode (negative) and a lead-dioxide electrode (positive). You can recharge them many times, as long as you take good care of them.

Some lead-acid batteries have semisolid electrolytes; they find applications in notebook computers and uninterruptible power supplies (UPSs). One of their greatest attributes is the fact you need not worry about the electrolyte spilling out if you tip them over or jostle them around.

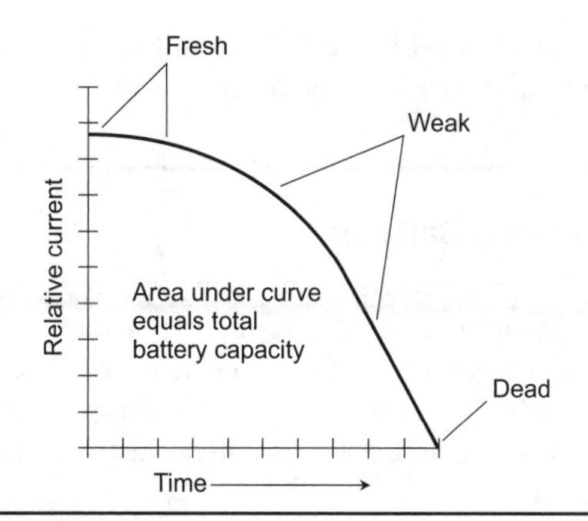

Figure 9-3 A declining discharge curve.

Wow!

A large lead-acid battery, such as the one in your vehicle, can store several tens of ampere-hours and deliver hundreds of amperes for a short time (a second or two).

Nickel-Based Cells and Batteries

Nickel-based cells include the *nickel-cadmium* (NICAD or NiCd) type and the *nickel-metal-hydride* (NiMH) type. *Nickel-based batteries* are available in packs of rechargeable cells. You can sometimes plug these packs directly into consumer equipment. In other cases, the batteries form part of the device housing, and the device itself keeps the batteries charged. You can put these cells and batteries through thousands of *charge/discharge cycles* if you take good care of them.

Nickel-based cells and batteries, particularly the NICAD type, sometimes exhibit a characteristic called *memory* or *memory drain*. If you use one repeatedly and allow it to discharge to the same extent with every cycle, it appears to lose most of its capacity. You can sometimes "cure" a nickel-based cell or battery of this problem by letting it run down almost all the way, recharging it, running it down again, and repeating the cycle a half dozen times before you use it again.

Nickel-based cells and batteries work best if used with charging units that take several hours to fully replenish the charge. So-called *high-rate* or *quick* chargers are available, but some of these can force too much current through a cell or battery. It's best if the charger is made especially for the cell or battery type you use.

In recent years, concern has mounted about the toxic environmental effects of the cadmium in NICAD cells and batteries. For this reason, NiMH cells and batteries have largely supplanted NICAD types for consumer use.

Tip

Never discharge a nickel-based battery until it "totally dies." If you do that, you can cause the polarity of one or more cells to reverse, ruining the battery for good.

Photovoltaic Cells and Batteries

A *photovoltaic* (PV) *cell* converts visible light, infrared (IR) rays, and/or ultraviolet (UV) rays directly into DC electricity. Figure 9-4 shows the basic construction of a silicon PV cell. The P-N junction forms the active region. The cell has a transparent housing so that radiant energy can strike the P-type silicon. Interconnected metal ribs form the positive electrode. The negative electrode is a metal backing in contact with N-type silicon.

Most silicon PV cells provide about 0.6 V DC in sunlight. If the current demand is low, muted daylight or artificial lamps can produce the full output voltage. As the

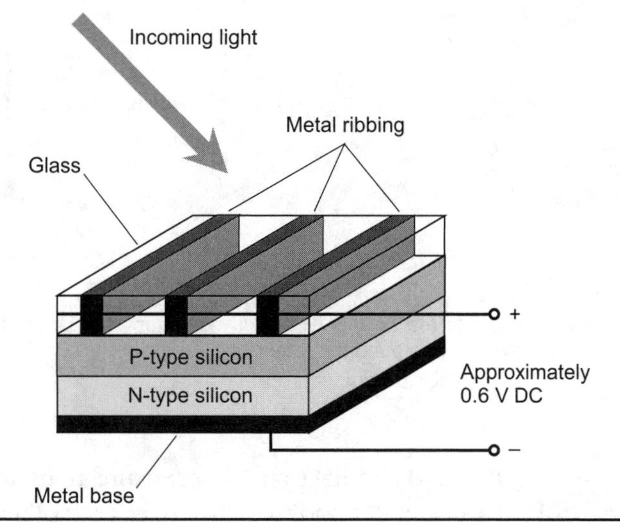

FIGURE 9-4 Construction of a silicon photovoltaic (PV) cell.

current demand increases, the cell must receive more intense illumination to produce full output voltage. The available current increases as the illuminance goes up, until you reach a state of *saturation* in which the cell delivers as much current as it can.

Wire Splicing

The simplest way to splice two single-conductor wires of the same diameter involves stripping the insulation off the ends for about an inch (2.5 centimeters) to expose the bare wire, bringing the bare-wire ends close together and parallel, and then wrapping them over each other to make a *twist splice*, as shown in Fig. 9-5. If the wires differ from each other in diameter, wrap the smaller wire around the larger one (Fig. 9-6). Once you've made the splice, wrap electrical tape over the connection.

When you want a wire splice to have significant mechanical strength, strip about 3 inches (7.5 centimeters) of insulation from the wire ends. Then bring the

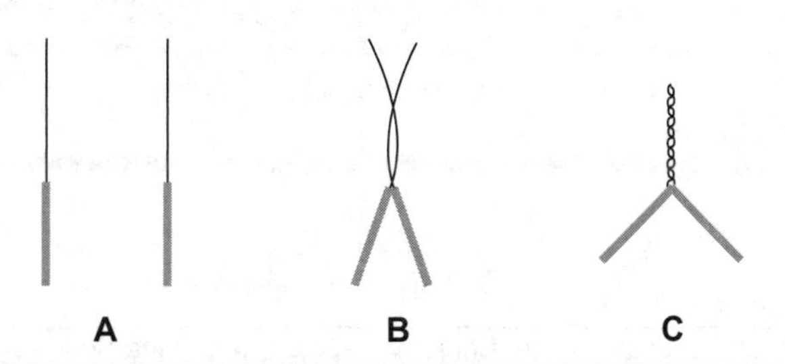

FIGURE 9-5 Twist splice for wires of the same diameter. Bring the exposed wire ends together and parallel (A), loop them around each other (B), and then wrap them over each other several times (C).

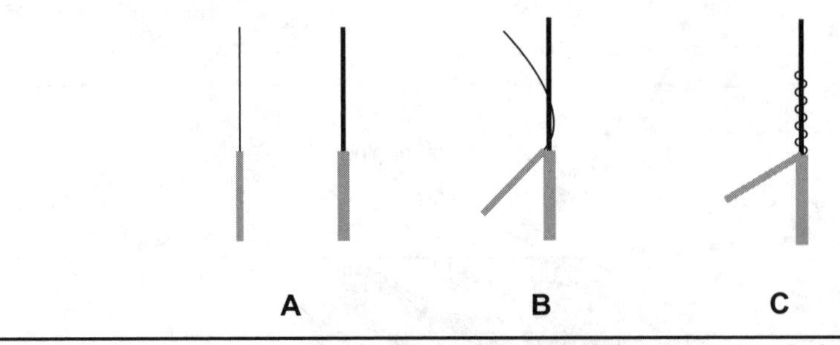

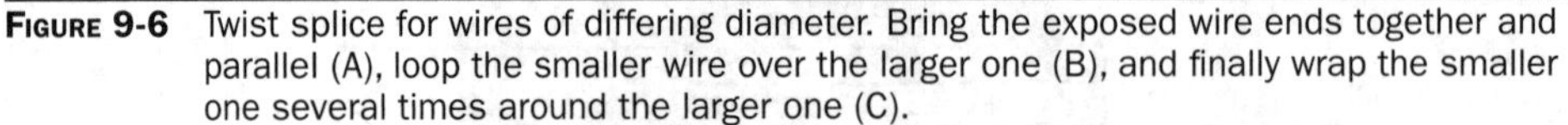

FIGURE 9-6 Twist splice for wires of differing diameter. Bring the exposed wire ends together and parallel (A), loop the smaller wire over the larger one (B), and finally wrap the smaller one several times around the larger one (C).

exposed wires together and parallel from opposite directions, overlapping them by about 2 inches (5 centimeters). Loop the wires over each other and then twist the exposed ends around each other several times on either side of the loop center, as shown in Fig. 9-7. Technicians call this type of connection a *Western Union splice.*

> **Tip**
>
> Apply solder to a splice if you need a solid, long-lasting electrical and mechanical bond. The soldering process provides extra tensile strength and keeps corrosion from degrading the electrical connection as time passes.

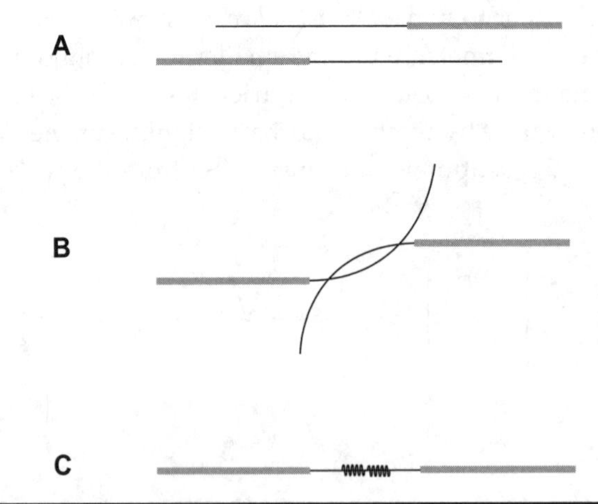

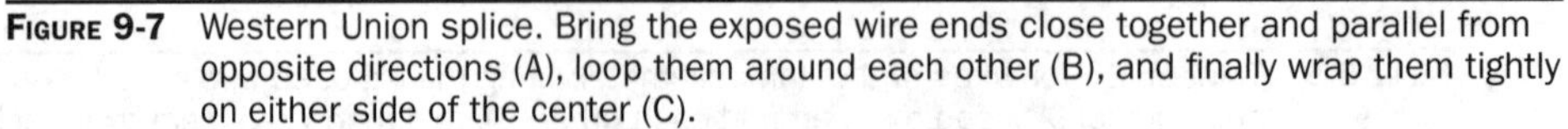

FIGURE 9-7 Western Union splice. Bring the exposed wire ends close together and parallel from opposite directions (A), loop them around each other (B), and finally wrap them tightly on either side of the center (C).

Soldering and Desoldering

Solder is a metal alloy used for securing electrical connections between conductors. There are several types of solder, intended for use with various metals and in diverse applications.

Types of Solder

The most common solder comprises tin and lead with a rosin core. Some types of solder have an acid core. In electronic devices, you should use rosin-core solder because acid-core solder will corrode the connections and can damage components. Acid-core solder is used for bonding sheet metal.

In tin/lead solder, the ratio of the constituent metals determines the temperature at which the solder will melt. In general, the higher the ratio of tin to lead, the lower the melting temperature. For general soldering purposes, you can use 50:50 solder. For heat-sensitive components, 60:40 or 63:47 solder works better because it melts at a lower temperature, allowing you to use a smaller soldering instrument.

Tin-lead solder is suitable for use with most metals except aluminum. *Aluminum solder* melts at a much higher temperature than tin-lead solder, and requires the use of a blowtorch or other high-heat device for application.

In high-current circuits, *silver solder* can withstand the high temperatures produced when large currents flow through components and connections. Tin-lead solder can melt in high-current environments, causing electrical and/or mechanical failure. You'll usually need a blowtorch for working with silver solder. This type of solder must be applied in a well-ventilated area because it produces dangerous fumes when heated.

> **Tip**
>
> Appendix K, in the back of this book, outlines the various types of solder most often used in electronics and industrial environments.

Soldering Instruments

A *soldering gun* is a quick-heating soldering tool. It gets its name from its general shape. You press a trigger-operated switch to heat the element up in a few seconds. Soldering guns are convenient in the assembly and repair of some kinds of electronic equipment. You'll find them in various wattage ratings for different applications.

A *soldering iron* consists of a heating element and a handle. The iron needs several minutes to fully heat after power-up; the larger the iron, the longer the warm-up time. Soldering irons are available in a wide range of wattages. The smallest are rated at a few watts and are used in miniaturized electronic equipment. The largest draw hundreds of watts and are used for outdoor wire splicing and sheet-metal bonding.

Soldering guns are more convenient than irons in situations in which you need moderately high heat and want to do point-to-point wiring. The soldering gun heats up and cools down rapidly. Small soldering irons are preferable for modifying and servicing miniaturized equipment, such as handheld radio transceivers and laptop computers.

For sheet-metal bonding and outdoor wire splicing, you can use a large soldering iron, but you might prefer a blowtorch. Its main advantage is portability; it does not require electricity, so you won't need a long, clumsy extension cord. You'll find small propane blowtorches in most hardware stores and large department stores.

> **Tip**
> A blowtorch is convenient for installing long-wire shortwave receiving antennas and wire amateur-radio antennas, as long as it's not raining or snowing, it's not too windy, and your area is not under a red-flag fire danger warning!

Printed Circuits

You'll usually do printed-circuit soldering from the non-component (foil) side of the board. Insert the component lead through the appropriate hole, and place the soldering iron so that it heats both the foil and the component lead (Fig. 9-8A). If the component is heat-sensitive, use a pair of needle-nosed pliers to grip the lead

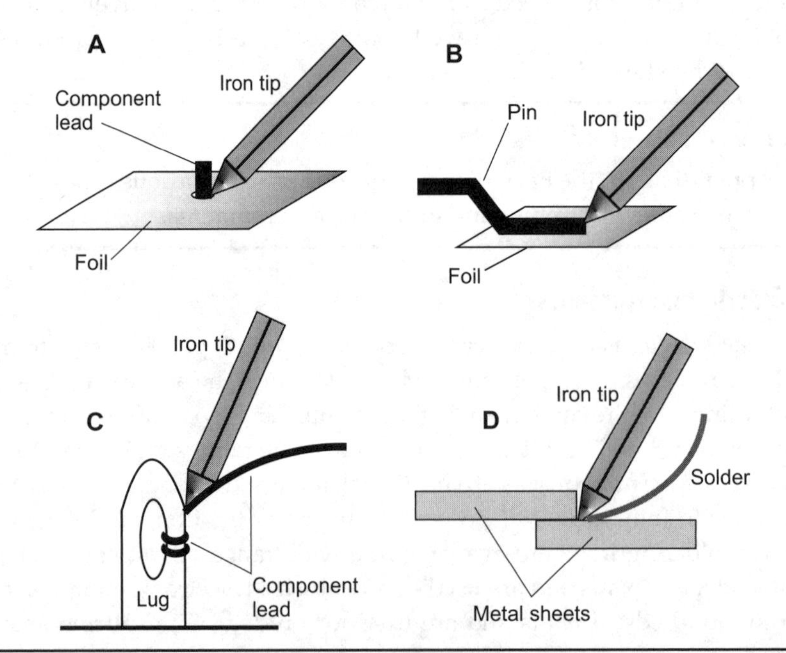

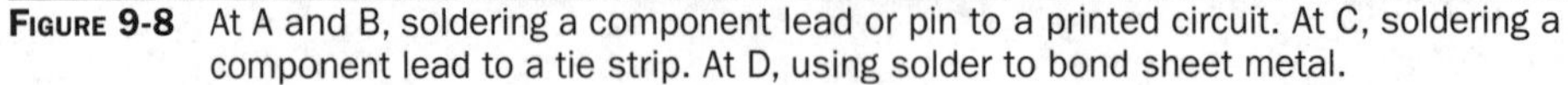
Figure 9-8 At A and B, soldering a component lead or pin to a printed circuit. At C, soldering a component lead to a tie strip. At D, using solder to bond sheet metal.

on the component (nonfoil) side of the board while you apply heat. Allow the solder to flow onto the foil and the component lead after the joint gets hot enough to melt the solder. Heating normally takes only a couple of seconds. The solder should completely cover the foil dot or square in which the component lead is centered. Don't use too much solder. After the joint has cooled, use a diagonal cutter to snip the component lead off flush with the solder.

If the circuit board has foil on both sides, it will have plated-through holes (or should have), and the soldering procedure described above will work okay. However, if the holes are not plated-through, you must apply solder to the foil and component lead on both sides of the board.

Some printed-circuit components are mounted on the foil side of the board. In such cases, *tin* (coat) the component lead and the circuit board foil with a thin layer of solder. Then place the component lead flat against the foil, and set the iron down in contact with the lead (Fig. 9-8B). The heat will melt the solder by conduction.

Point-to-Point

Most point-to-point wiring is accomplished by means of *tie strips*, also called *terminal strips*, in which one or more wires terminate.

When you're tie-strip wiring, tin the lug with a thin layer of solder. Don't do any soldering until all wires have been attached to the lug. Wrap each wire two or three times around the lug using a pair of needle-nosed pliers. Cut off excess wire using a diagonal cutter. When you've attached all wires to the lug, hold the soldering instrument against each wire "coil," one at a time, and heat up the connection until the solder flows freely between the wire turns, adhering to both the wire and the lug (Fig. 9-8C). Use enough solder to completely cover the connection, but don't let any solder ball up or drip from the connection.

Tip

If you must wire a heat-sensitive component to a tie strip, you can use a pair of needle-nosed pliers to conduct heat away from the component. Clamp the pliers on the component lead between the connection and the component body. Don't remove the pliers until the soldered joint has cooled down almost to room temperature.

Don't Get "Cold"!

If you fail to apply enough heat to a solder connection, you'll end up with a so-called *cold solder joint*. A properly soldered connection looks shiny and clean; a cold joint looks dull or rough. Many electronic equipment failures occur because of cold solder joints, which can exhibit high resistance and/or intermittent conduction. If you ever find a cold joint, remove as much of the solder as possible using wire braid called *solder wick* that's designed especially for desoldering. Clean the surfaces and then resolder the connection.

Sheet Metal Soldering

When you must solder sheet metal other than aluminum, use rosin-core solder if possible. Once in a while, rosin won't allow a good enough mechanical bond. In such cases, you can use acid-core solder. Special solder is available for use with aluminum, which does not readily adhere to most other types of solder.

Use fine emery paper to sand down the sheet-metal surfaces along the edges that you want to bond. Then clean the surfaces with a non-corrosive, grease-free solvent, such as isopropyl alcohol. Use a high-wattage soldering iron or blowtorch to heat the metal while you tin both surfaces with a thin layer of solder. Then secure the sheets in place and apply heat to the metal. Heat them up until the solder completely melts where the surfaces make contact. Slowly apply additional solder on each side of the bond, working gradually along the length of the bond from one end to the other (Fig. 9-8D). The bond will require some time to cool. Keep it free from mechanical stress until it has cooled completely.

Warning! **Don't apply water to a hot solder joint or sheet-metal bond to hasten the cooling process. It'll cool the region down in a hurry, but it can produce a brittle bond. In addition, a little bit of water, when suddenly heated, might boil off and injure you. One drop of scalding water in your eye can propel you straight to your hospital's emergency room!**

Removing Solder

When you need to replace a faulty component, you'll probably have to desolder one or more connections. With most printed-circuit boards, desoldering involves the steady application of heat as you "suck away" the solder from the connection with solder wick. Heat the connection and the wick to a temperature sufficient to melt the solder. Apply only enough heat to ensure that the connection gets desoldered without damaging the circuit board and nearby components.

Diverse desoldering aids exist on the market today. One popular device employs an air-suction nozzle, which swallows the solder by vacuum action as a soldering iron heats the connection. This apparatus works fast, and you'll find it handy when you need to desolder a lot of connections. It's also useful in desoldering tiny connections, in which little room for error exists. For large connections such as wire splices and solder-welded joints, you'll probably do better to remove the entire connection than to try to desolder it.

Cords and Cables

You'll encounter a vast variety of cables for the transmission of electrical power, RF, and data signals over short to moderate spans. Here's a look at the most common cable configurations.

Lamp Cord

The simplest electrical cable, other than a single wire, is two-conductor *lamp cord,* also known as *zip cord.* It works well with common appliances at low to moderate current levels. Two wires are embedded in rubber or plastic insulation that serves as the jacket (Fig. 9-9A). The individual wires are stranded to help them resist breakage from repeated flexing.

> **Of Course, You Knew**
>
> Some appliance cords, especially those for heavy-duty household items such as electric toasters and skillets, have three conductors. The third wire facilitates grounding to protect you from electrocution if the device partially shorts out (not enough to trip a breaker or blow a fuse, but enough to endanger you!).

Multiconductor Cable

When a cable has several wires, they can be individually insulated, bundled together, and enclosed in a tough external jacket, as shown in Fig. 9-9B. If the cable must have

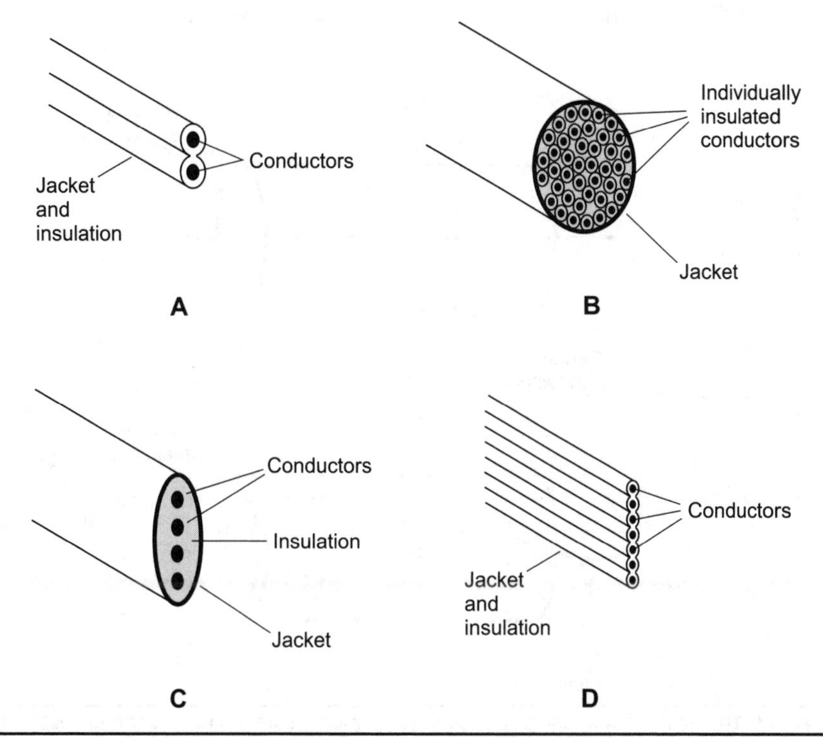

FIGURE 9-9 Two-wire lamp cord (A), bundled multiconductor cable (B), flat multiconductor cable (C), and multiconductor ribbon (D).

good flexibility, each wire is stranded. If only a few conductors exist, they can run parallel to each other and are protected by a tough plastic outer jacket, as shown in Fig. 9-9C, a configuration called *flat cable*. Sometimes, several conductors are molded into a flexible, hard, thin plastic strip, as shown in Fig. 9-9D, an arrangement known as *multiconductor ribbon*.

Tip

Look for multiconductor ribbon inside high-tech electronic devices, particularly computers. It's physically sturdy and takes up a minimum of space.

Coaxial Cable

Coaxial cable, also called *coax* (pronounced "*co*-ax"), serves as the transmission medium of choice for RF signal transmission because of its excellent shielding against electromagnetic (EM) fields at all frequencies.

In a typical coaxial cable, a cylindrical shield surrounds a single center conductor. In some cable types, solid or foamed polyethylene insulation, called the *dielectric*, keeps the inner conductor running exactly down the center of the cable (Fig. 9-10A). Other cables have a thin tubular layer of solid polyethylene just inside the shield (Fig. 9-10B), so air makes up most of the dielectric medium.

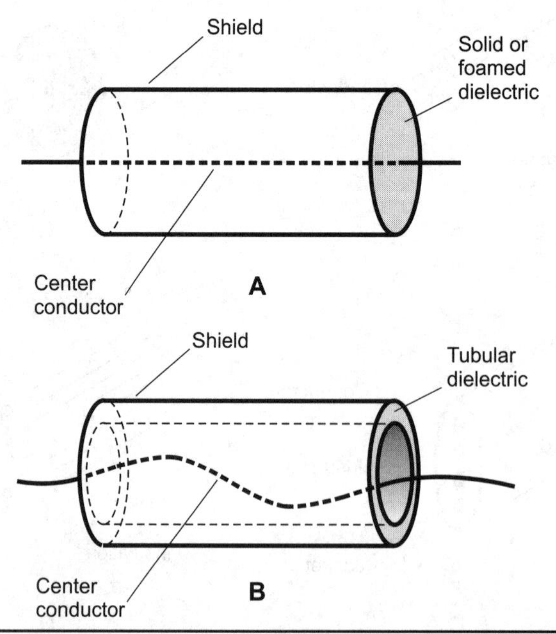

Figure 9-10 At A, coaxial cable with solid or foamed insulation (called the dielectric) between the center conductor and the shield. At B, similar cable in which air makes up most of the dielectric medium.

The most sophisticated coaxial cables have a solid metal shield that resembles a pipe or conduit. Engineers call it *hardline*. It's available in larger diameters than most coaxial cables are, and it transmits signals with minimal loss. You'll find hardline in high-power, fixed radio and television transmitting installations.

Cable Splicing

When you want to splice cord, multiconductor cable, or ribbon cable, you'll do best if you use a Western Union splice for each individual conductor and solder each connection. Wrap all splices individually with electrical tape, and then wrap the entire combination with electrical tape afterwards. Insulating the individual connections ensures that no two conductors can come into electrical contact with each other inside the splice. Insulating the whole combination keeps the cable from shorting out to external metallic objects, and provides additional protection against corrosion and oxidation.

You can twist-splice two-conductor lamp cord or ribbon cable. First, bring the ends of the cables together and pull the conductors outward from the center, as shown in Fig. 9-11A. Then twist the corresponding conductor pairs several times around each other, as shown in Fig. 9-11B. Solder both connections and then trim the splices to lengths of approximately 1/2 inch (1.2 centimeters). Insulate each twist splice carefully with electrical tape. Then fold the twists back parallel to the cable axis in opposite directions. Finally, wrap the entire connection with electrical tape to insulate it and hold the splices in place.

> **Tip**
>
> To minimize the risk of a short circuit between conductors in a multiconductor cable, position the splices for each conductor at slightly different points along the cable.

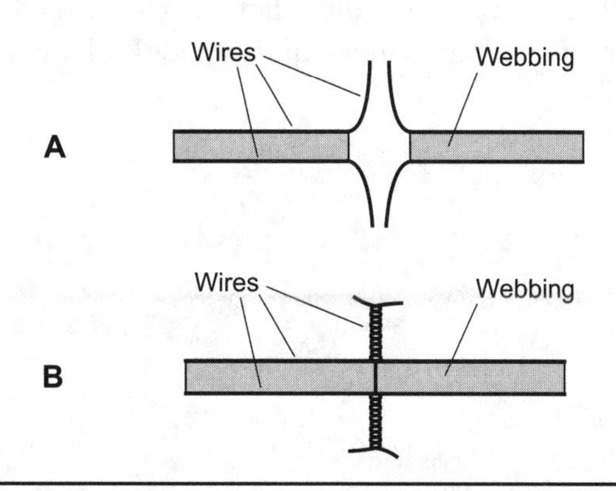

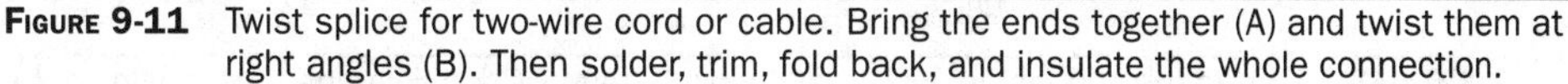

FIGURE 9-11 Twist splice for two-wire cord or cable. Bring the ends together (A) and twist them at right angles (B). Then solder, trim, fold back, and insulate the whole connection.

Plugs and Connectors

Plugs attached to the ends of appliance cords have two or three prongs that fit into *outlets* having receptacles in the same configuration. The plugs are called *male connectors*, and the receptacles into which the plugs fit are called *female connectors*.

Appliance plugs and outlets provide temporary electrical connections. They can't offer reliability in long-term applications because the prongs on any plug, and the metallic contacts inside any receptacle, will eventually oxidize or corrode. A new male plug has shiny metal blades, but an old one has visibly tarnished blades. Similar oxidation occurs in the "slots" of a female receptacle, even though you can't see it.

A *clip lead* comprises a short length of flexible wire, equipped at one or both ends with a simple, temporary connector. Clip leads do a poor job in permanent installations, especially outdoors, because corrosion occurs easily, and the connector can slip out of position. In addition, clip-lead connections can't carry much current. Clip leads are used primarily in DC and low-frequency AC applications at low voltage and current levels.

A *banana connector* is a convenient single-lead connector that slips easily in and out of its receptacle. The "business part" of a *banana plug* (male connector) looks like a round peg, as shown in Fig. 9-12. *Banana jacks* (female receptacles, not shown in the figure) are round holes. You'll sometimes find banana jacks inside the screw terminals of low-voltage DC power supplies.

Warning! **Banana connectors, like clip leads, are designed for low-voltage, low-current, and short-term indoor use only. Never employ these connectors in high-voltage applications. If you do, the exposed conductors will pose a shock hazard.**

A *hermaphroditic connector* is an electrical plug/jack with two or more contacts, some of them male and some of them female. Usually, hermaphroditic connectors at opposite ends of a single length of cable look identical when viewed face-on.

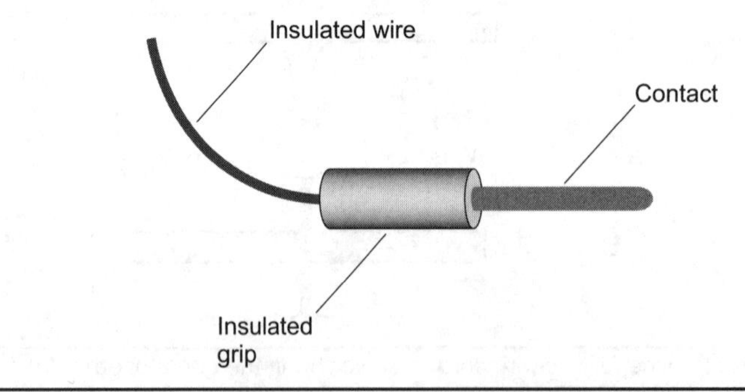

FIGURE 9-12 Banana connectors work well in low-voltage DC applications. The single contact (the plug, shown here) slides into a cylindrical receptacle (the jack).

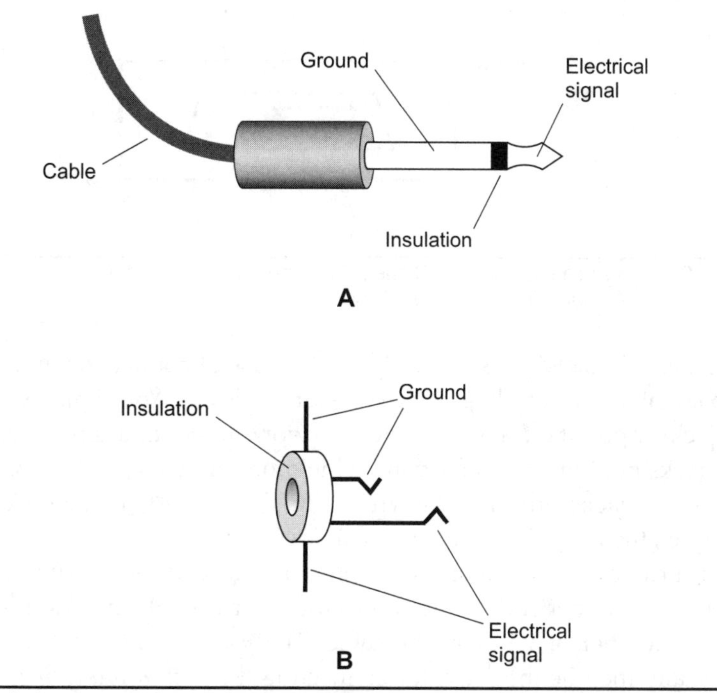

FIGURE 9-13 At A, a two-conductor phone plug. At B, a two-conductor phone jack.

However, the pins and holes have asymmetrical geometry, so you can join the two connectors in the correct way only. This "can't-go-wrong" feature makes hermaphroditic connectors ideal for polarized circuits, such as DC power supplies, and for multiconductor electrical control cables.

Phone plugs and jacks find extensive use in DC and low-frequency AC systems at low-voltage current levels. In its conventional form, the male phone plug (Fig. 9-13A) has a rod-shaped metal *sleeve* that serves as one contact, and a spear-shaped metal *tip* that serves as the other contact. A ring of hard-plastic insulation separates the sleeve and the tip. Typical diameters are 1/8 inch (3.175 millimeters) and 1/4 inch (6.35 millimeters). The female phone jack (Fig. 9-13B) has contacts that mate securely with the male plug contacts. The female contacts have built-in spring action that holds the male connector in place after insertion.

Engineers originally designed the phone plug and jack for use with two-conductor cables. In recent decades, three-conductor phone plugs and jacks have become common as well. You'll find them in high-fidelity stereo sound systems and in the audio circuits of multimedia computers and radio receivers. The male plug has a sleeve broken into two parts along with a tip, and the female connector has an extra contact that touches the second sleeve when you insert the plug into the jack. You'll even see four-conductor phone plugs and jacks used with microphone/headphone combinations for smartphones and tablet computers.

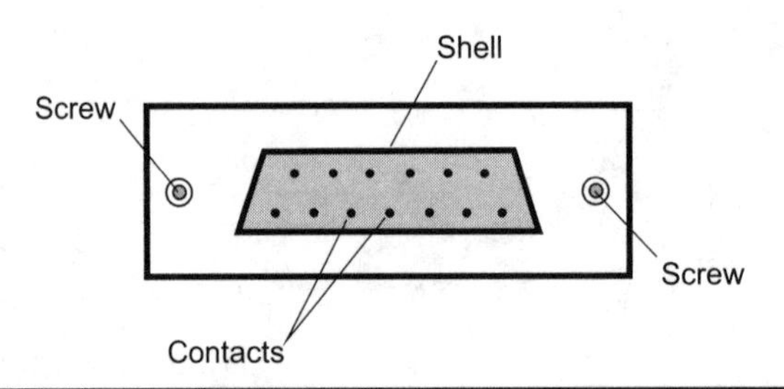

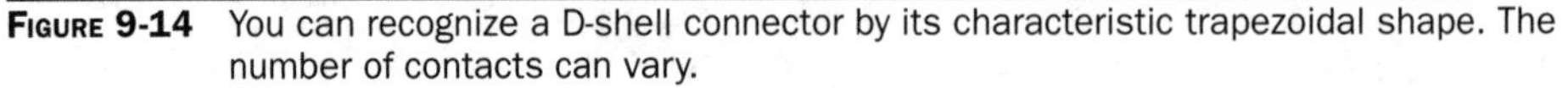
FIGURE 9-14 You can recognize a D-shell connector by its characteristic trapezoidal shape. The number of contacts can vary.

Phono plugs and jacks are designed for ease of connection and disconnection of coaxial cable at low voltages and low-current levels. You simply push the plug onto the jack, or pull it off. These connectors work in the same situations as phone plugs and jacks do, but they offer better shielding for coaxial-cable connections. Phono plugs and jacks are also known as *RCA connectors*, named after their original designer, the *Radio Corporation of America* (RCA).

If a cable contains more than three or four conductors, you can put a *D-shell connector* on either end. These connectors are available in various sizes, depending on the number of wires in the cable. The *ports* on an older personal computer, especially the one intended for connecting the *central processing unit* (CPU) to an external *video display*, commonly employ D-shell connectors, which have the characteristic appearance shown in Fig. 9-14. The trapezoidal shell forces you to insert the plug correctly; you can't accidentally put it in upside-down. The female socket has holes into which the pins of the male plug slide. Screws or clips secure the plug once it's in place.

Tip

You can remove an oxide coating on connector contacts by rubbing them with fine-grain sandpaper, emery paper, or steel wool, and then wiping the contacts off with a dry cloth. Continue the cleaning process until bright metal shows everywhere on the exposed parts of the contacts. If the contact prongs are too small or too closely spaced for this cleaning method to work, you can use specially formulated contact cleaner to remove the oxidation layer. You can find this type of cleaner sold in good hardware or electronics stores.

Oscilloscope

An *oscilloscope* is a lab test instrument that displays signal waveforms. The display appears as a graph. The screen is calibrated in a grid pattern for reference. As you gain experience, you'll eventually want one. The best, new ones have liquid crystal displays (LCDs). However, you might prefer to buy an older cathode-ray-tube (CRT) type on the used market and save a few dollars!

The Display

Figure 9-15 shows a typical oscilloscope screen with an AC sine wave displayed. An oscilloscope can provide an indication of whether a waveform has the desired shape, and whether it has the correct amplitude, both positively and negatively, to work in a given circuit or application.

> **Tip**
>
> You can use an oscilloscope to measure frequency, but not with great accuracy. A frequency counter (described below) works better for that purpose.

When no signal exists at the input, the beam sweeps horizontally across the screen in a straight line. In the most common oscilloscope configuration, known as the *time domain*, the horizontal axis shows time. You can adjust the *sweep rate* (speed of the trace on the screen) in seconds, milliseconds, microseconds, or nanoseconds per division. A division measures about one centimeter (1 cm) wide or tall.

When you input an AC signal to the scope, the trace deflects vertically up and down. The extent of the movement up or down from moment to moment depends on the instantaneous positive or negative amplitude of the signal. You can adjust the display sensitivity as portrayed in volts, millivolts, microvolts, or nanovolts per division.

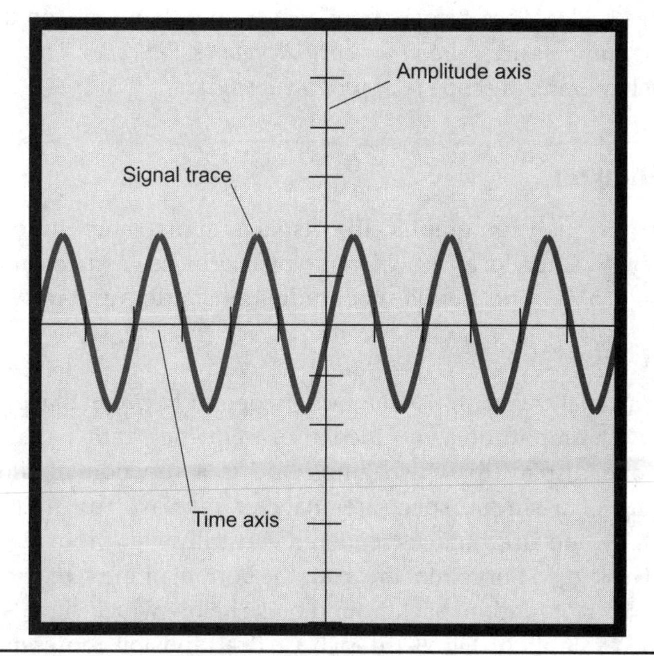

FIGURE 9-15 An oscilloscope display. The horizontal axis portrays time. The vertical axis renders positive and negative signal voltage.

Tip

Some oscilloscopes have signal inputs for both the horizontal and vertical deflections. This feature allows comparison of two signals for frequency and phase. Engineers call the resulting patterns *Lissajous figures*. Google on that term!

Other Common Features

One of the most important requirements of an oscilloscope is that the display remain stable at all times. To make this happen, the sweep must be exactly synchronized with the frequency, so that the trace goes over the same places on the screen with each pass. This coordination is ensured by *triggering*, in which the sweep rate locks into the frequency of the input signal. Most scopes can be triggered based on either the input signal or some external signal source.

Most oscilloscopes can reveal waveform details for signals having frequencies of up to several megahertz. Special wideband oscilloscopes exist that will resolve signal details at much higher frequencies.

Sometimes you'll want to compare two waveforms that have the same frequency, or that are harmonically related. A *dual-trace oscilloscope* does this by displaying a separate trace for each signal, one above the other.

A *persistence-trace oscilloscope* holds a display for several seconds or minutes after you remove the signal. A *storage oscilloscope* can store and recall waveforms, in much the same way that a memory calculator works.

Some oscilloscopes display frequency, rather than time, on the horizontal axis. These instruments are called *frequency-domain oscilloscopes*. The most common type of frequency-domain scope is a *spectrum analyzer*.

Spectrum Analyzer

A spectrum analyzer graphically displays signal amplitude as a function of frequency. Eventually, as a hobbyist, you might use a spectrum analyzer to help you design, align, and troubleshoot radio transmitters and receivers.

Concept

A spectrum analyzer comprises an oscilloscope and a circuit that provides the spectral display with amplitude as a function of frequency, rather than the conventional oscilloscope display in which amplitude is a function of time. Figure 9-16 is a drawing of a simple spectrum-analyzer display. The frequency is rendered horizontally, and amplitude is rendered vertically. Here, four signals appear on the screen. The jagged horizontal line near the bottom of the screen represents the *noise floor*: the level of random background noise below which signals can't show up.

In the example of Fig. 9-16, each vertical division represents 10 dB of signal level change, which corresponds to a 10:1 power ratio. The 0 dB level near the top

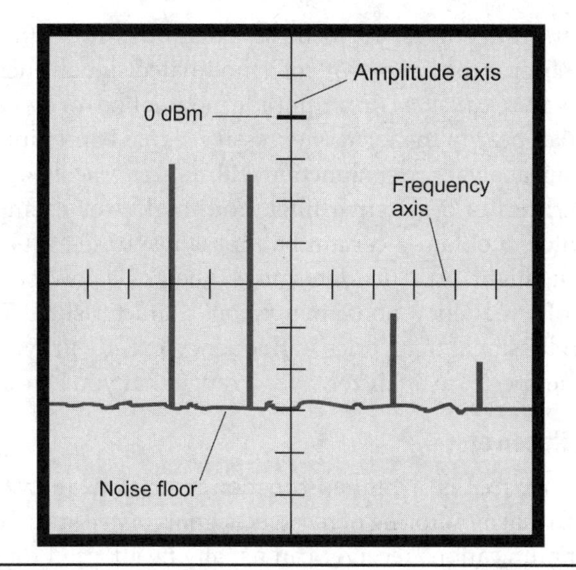

FIGURE 9-16 A spectrum analyzer display. The horizontal axis portrays frequency. The vertical axis renders signal strength.

of the screen serves as a reference point for comparison of amplitudes below it. A typical level is 1 milliwatt (1 mW) for the 0 dB point. Engineers call it 0 dBm (0 dB with respect to 1 mW). Each horizontal division represents 10 kHz of frequency change. You can adjust these parameters at will. You can also set the amplitude display for a linear scale (say, 0.1 volt per division) if you prefer, rather than the more common logarithmic (decibel) scale.

Some spectrum analyzers are designed for the AF range. These include narrowband analyzers of fixed bandwidth, such as 5 Hz, and devices having a bandwidth that's a constant percentage of the frequency of the signal being evaluated. Examples of the latter are octave, half-octave, and 1/3-octave analyzers. Real-time analyzers commonly have visual displays of response in 1/3-octave bands covering the audio range.

Operation

To use a spectrum analyzer, you first choose, by means of a selector switch, the band of frequencies that you want to examine. This band might cover a vast swath of radio spectrum from direct current to 1 GHz or more, or it might cover a small frequency range, say 10 kHz wide. You then adjust the gain (vertical-scale sensitivity) as desired. You must also adjust the resolution and sweep rate to suit the intended application.

You can use a spectrum analyzer to determine the levels of harmonics and spurious emissions coming from a radio transmitter. In the United States, radio equipment must meet certain government-imposed standards regarding the purity of emissions. A spectrum analyzer gives you an immediate indication of whether or not a transmitter lives up to these standards. You can also use a spectrum analyzer to observe the bandwidth of a modulated signal. Such improper operating conditions as *overmodulation* (in amplitude modulation) or *overdeviation* (in frequency modulation) appear in the form of excessive signal bandwidth.

A spectrum analyzer, in conjunction with a *sweep generator,* can be used to evaluate the characteristics of a selective filter. You might, for example, need to adjust a bandpass filter to obtain a certain bandpass response at the front end of a radio receiver. You might want to determine whether a low-pass filter provides the desired cutoff frequency and attenuation characteristics. The sweep generator produces an RF signal that varies in frequency, exactly in synchronization with the display of the spectrum analyzer, over a range that you can select.

Panoramic Receiver

A *panoramic receiver* is a limited-function spectrum analyzer adapted to allow continuous visual monitoring of received signals over a specific band of frequencies. A radio communications receiver can usually be adapted for panoramic reception by connecting a spectrum analyzer into the intermediate-frequency (IF) chain.

In the panoramic receiver, a display shows signals as vertical pips along a horizontal axis. The signal amplitude is indicated by the height of the pip. The position of the pip along the horizontal axis indicates its frequency. The frequency to which the receiver is tuned appears at the center of the horizontal scale. You can set the frequency increment per horizontal division for spectral analysis of a single signal (for example, 0.5 kHz per division) or observation of a significant frequency span (for example, 50 kHz per division). The maximum possible frequency display range is limited by the characteristics of the receiver's IF stages. You can set the amplitude change per vertical division to a certain number of decibels (usually 3 dB or 10 dB), or you can set it for a linear indication.

> **Tip**
>
> Some advanced communications receivers have panoramic reception built-in. Display quality varies from rotten to excellent. The most expensive high-end receivers have panoramic reception displays that rival those of dedicated spectrum analyzers.

Frequency Counter

A *frequency counter* measures and displays the frequency of a signal by counting the cycles or pulses that occur within a specific, constant period of time called the *gate time.* You'll find the direct digital readout easy to interpret. Typical frequency counters have readouts that show six to ten significant digits. The gate defines the starting and ending time for each counting period or "window." The longer the gate time, the more precise the measurement.

You'll need an accurate time standard for a counter to work at its best. You can use a crystal oscillator as the *clock* ("metronome") for the counter. You can synchronize the clock frequency with an accepted time standard, such as the WWV/WWVH shortwave radio broadcasts. These time standards can also function directly as counting clocks. The stations transmit "ticks" at 1-second intervals. Frequency dividers and/or multipliers can convert them to any interval you want.

Signal Generator

A *signal generator* is an oscillator, often equipped with a modulator, that you can use in sophisticated tests of audio hi-fi or RF communications equipment. Most signal generators are intended for either AF or RF applications, but not both.

In its simplest form, a signal generator consists of an oscillator that produces a sine wave of a certain amplitude in microvolts or millivolts, and a certain frequency in hertz, kilohertz, or megahertz. Some AF signal generators can produce several different types of waveforms including square, sawtooth, ramp, and triangular. The most sophisticated signal generators for RF testing have amplitude modulators and/or frequency modulators built-in.

APPENDIX A

Schematic Symbols

ammeter

amplifier, general

amplifier, inverting

amplifier, operational

AND gate

antenna, balanced

antenna, general

antenna, loop

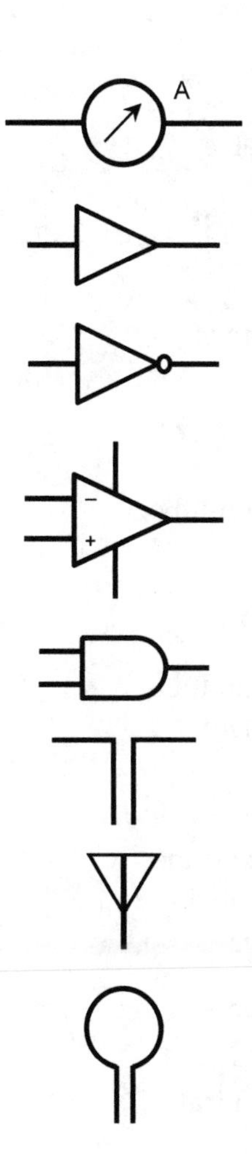

antenna, loop, multiturn

battery, electrochemical

capacitor, feedthrough

capacitor, fixed

capacitor, variable

capacitor, variable,
split-rotor

capacitor, variable,
split-stator

cathode, electron-tube,
cold

cathode, electron-tube,
directly heated

cathode, electron-tube,
indirectly heated

cavity resonator

cell, electrochemical

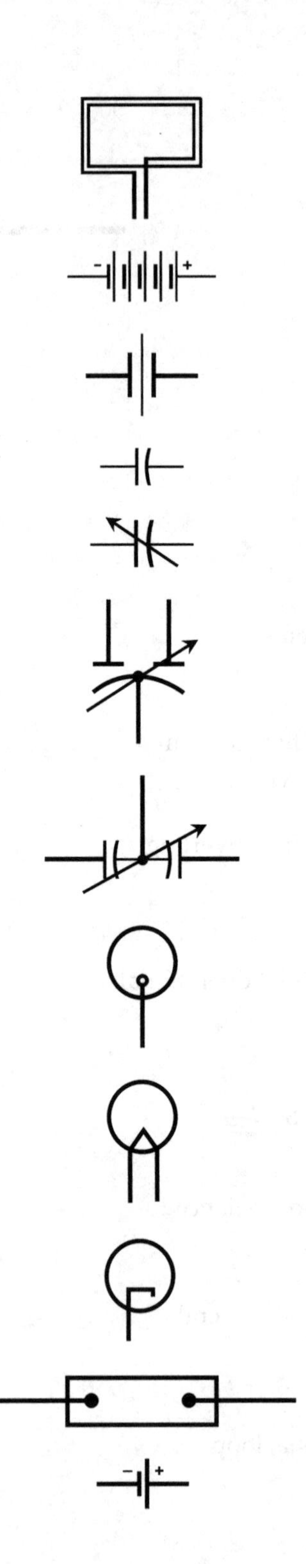

circuit breaker

coaxial cable

crystal, piezoelectric

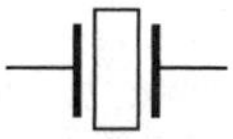

delay line

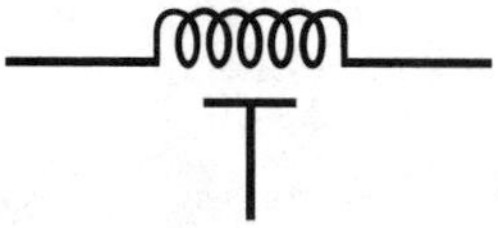

diac

diode, field-effect

diode, general

diode, Gunn

diode, light-emitting

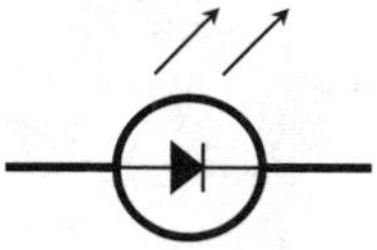

diode, photosensitive

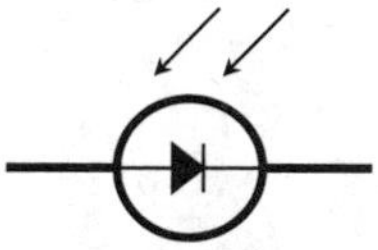

diode, PIN

diode, Schottky

diode, tunnel

diode, varactor

diode, Zener

directional coupler

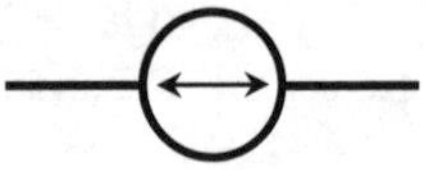

directional wattmeter

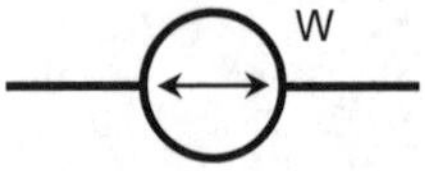

exclusive-OR gate

female contact, general

Ferrite bead

filament, electron-tube

fuse

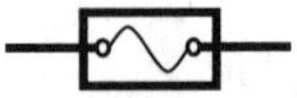

galvanometer

grid, electron-tube

ground, chassis

ground, earth

handset

headset, double

headset, single

headset, stereo

inductor, air core

inductor, air core, bifilar

inductor, air core, tapped

inductor, air core, variable

inductor, iron core

inductor, iron core, bifilar

inductor, iron core, tapped

inductor, iron core, variable

inductor, powdered-iron core

inductor, powdered-iron core, bifilar

inductor, powdered-iron core, tapped

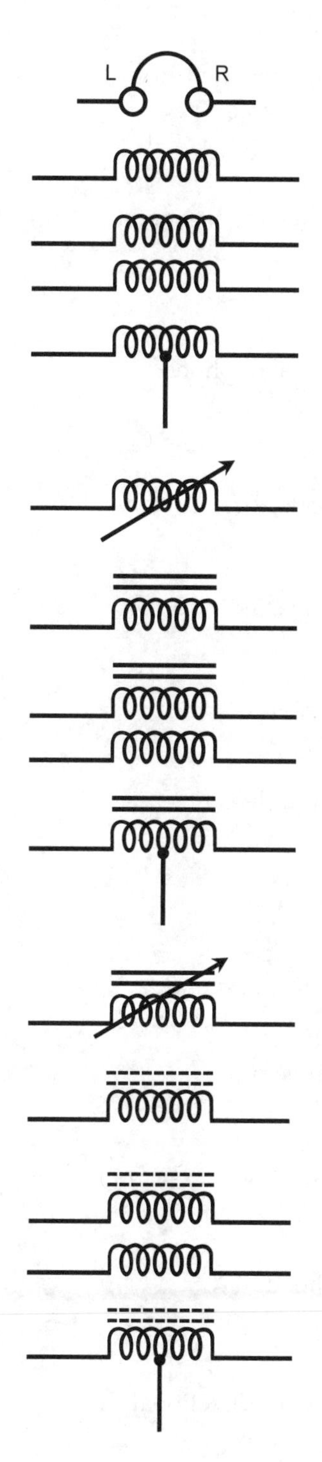

inductor, powdered-iron core, variable

integrated circuit, general

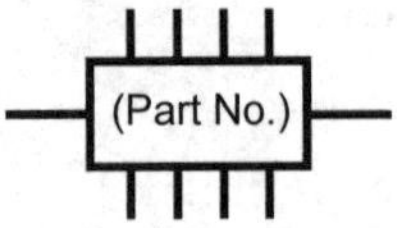

jack, coaxial or phono

jack, phone, 2-conductor

jack, phone, 3-conductor

key, telegraph

lamp, incandescent

lamp, neon

male contact, general

meter, general

microammeter

microphone

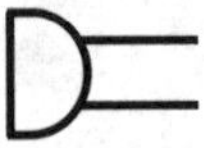

microphone, directional

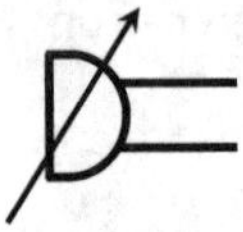

milliammeter

NAND gate

negative voltage connection

NOR gate

NOT gate

optoisolator

OR gate

outlet, 2-wire, nonpolarized

outlet, 2-wire, polarized

outlet, 3-wire

outlet, 234-volt

plate, electron-tube

plug, 2-wire, nonpolarized

plug, 2-wire, polarized

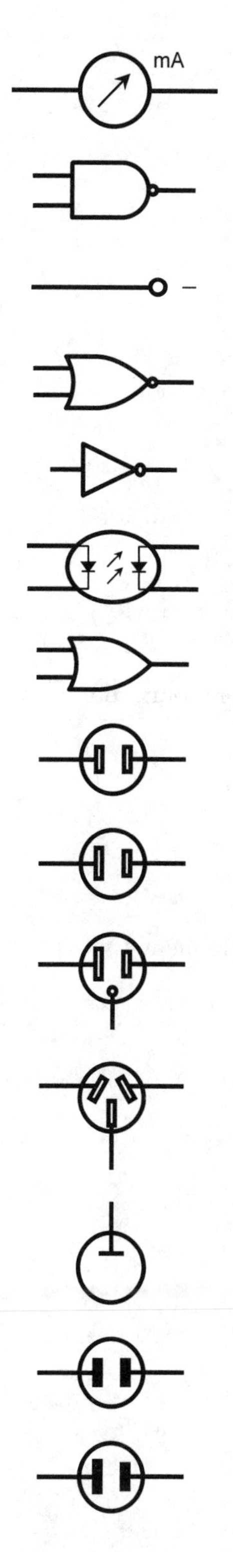

plug, 3-wire

plug, 234-volt

plug, coaxial or phono

plug, phone, 2-conductor

plug, phone, 3-conductor

positive voltage connection

potentiometer

probe, radio-frequency

rectifier, gas-filled

rectifier, high-vacuum

rectifier, semiconductor

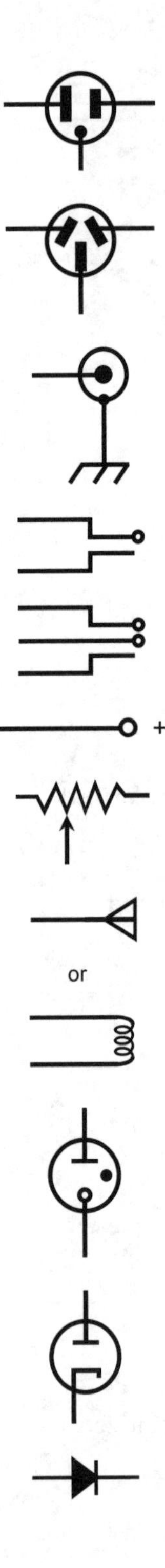

rectifier, silicon-controlled

relay, double-pole, double-throw

relay, double-pole, single-throw

relay, single-pole, double-throw

relay, single-pole, single-throw

resistor, fixed

resistor, preset

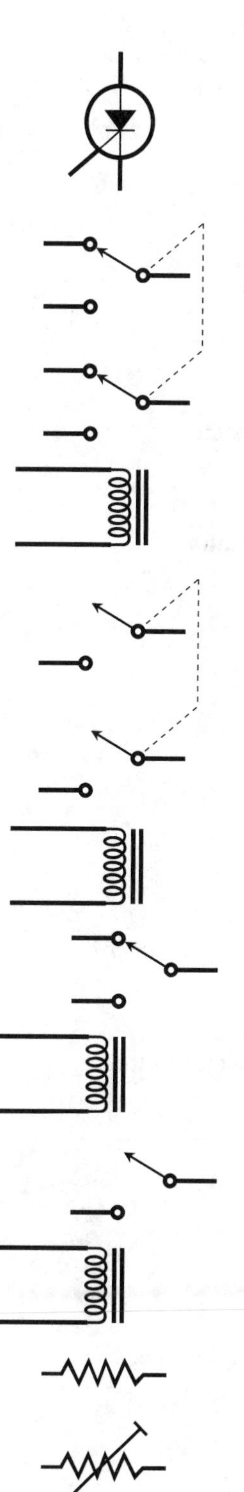

resistor, tapped

resonator

rheostat

saturable reactor

signal generator

solar battery

solar cell

source, constant-current

source, constant-voltage

speaker

switch, double-pole,
double-throw

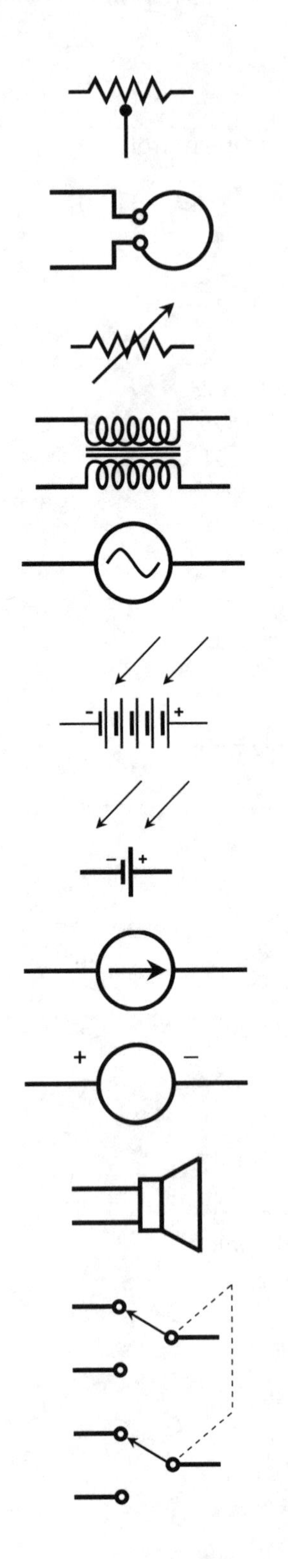

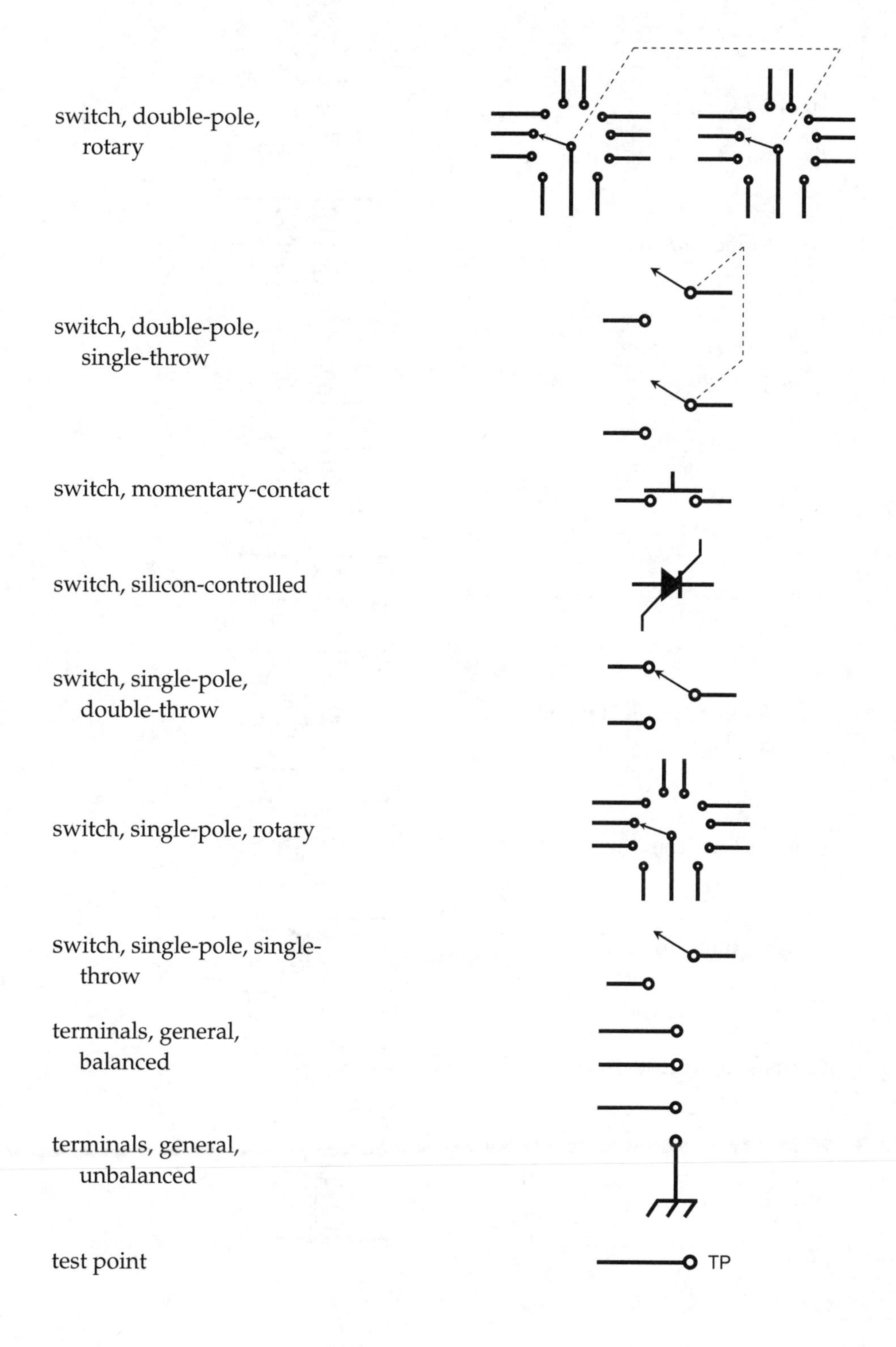
switch, double-pole, rotary
switch, double-pole, single-throw
switch, momentary-contact
switch, silicon-controlled
switch, single-pole, double-throw
switch, single-pole, rotary
switch, single-pole, single-throw
terminals, general, balanced
terminals, general, unbalanced
test point
TP

thermocouple

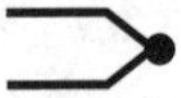

or

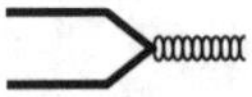

transformer, air core

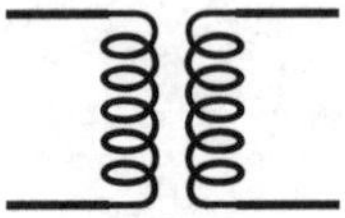

transformer, air core, step-down

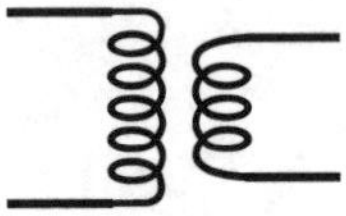

transformer, air core, step-up

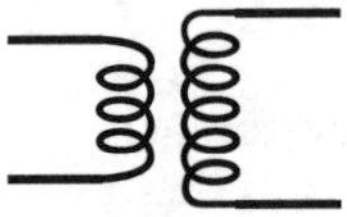

transformer, air core, tapped primary

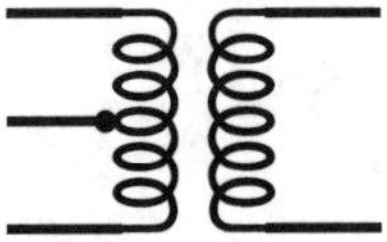

transformer, air core, tapped secondary

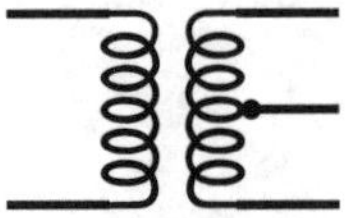

transformer, iron core

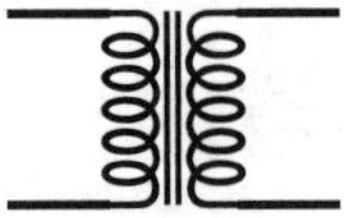

transformer, iron core, step-down

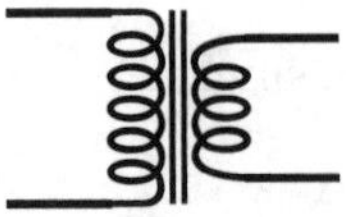

transformer, iron core, step-up

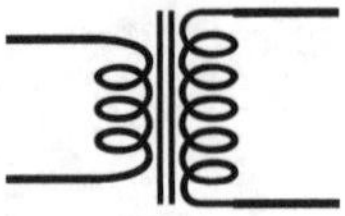

transformer, iron core, tapped primary

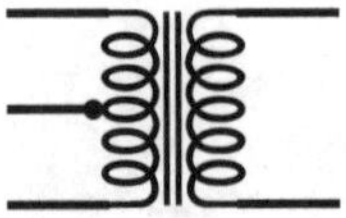

transformer, iron core, tapped secondary

transformer, powdered-iron core

transformer, powdered-iron core, step-down

transformer, powdered-iron core, step-up

transformer, powdered-iron core, tapped primary

transformer, powdered-iron core, tapped secondary

transistor, bipolar, NPN

transistor, bipolar, PNP

transistor, field-effect, N-channel

transistor, field-effect, P-channel

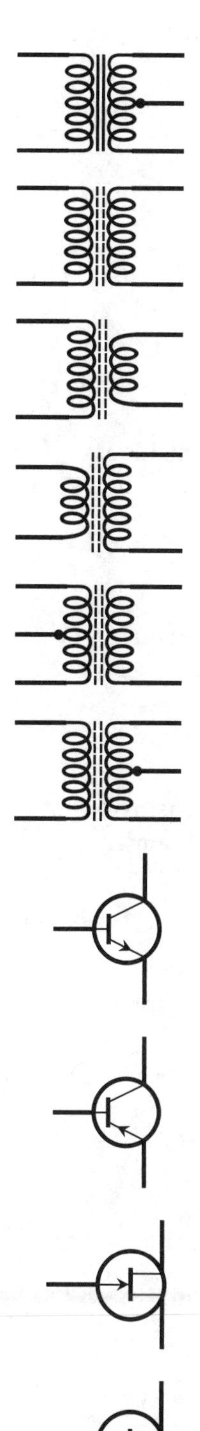

transistor, MOS field-effect,
N-channel

transistor, MOS field-effect,
P-channel

transistor, photosensitive,
NPN

transistor, photosensitive,
PNP

transistor, photosensitive,
field-effect, N-channel

transistor, photosensitive,
field-effect, P-channel

transistor, unijunction

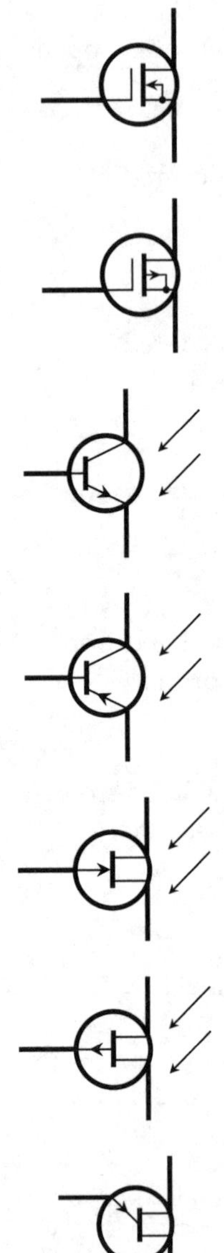

triac

tube, diode

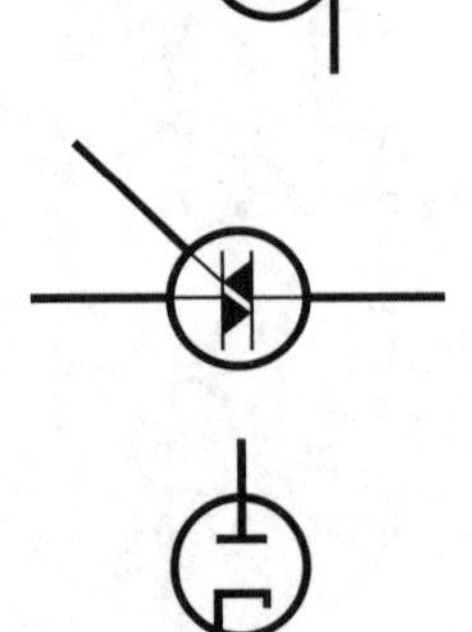

tube, heptode

tube, hexode

tube, pentode

tube, photosensitive

tube, tetrode

tube, triode

unspecified unit or
component

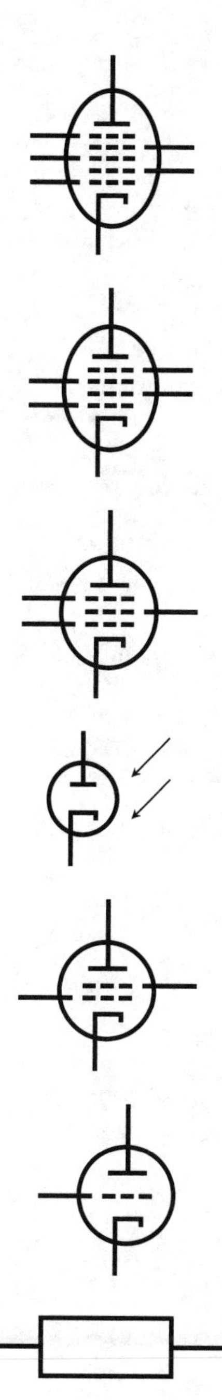

voltmeter

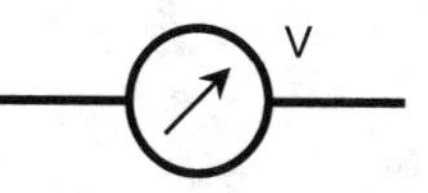

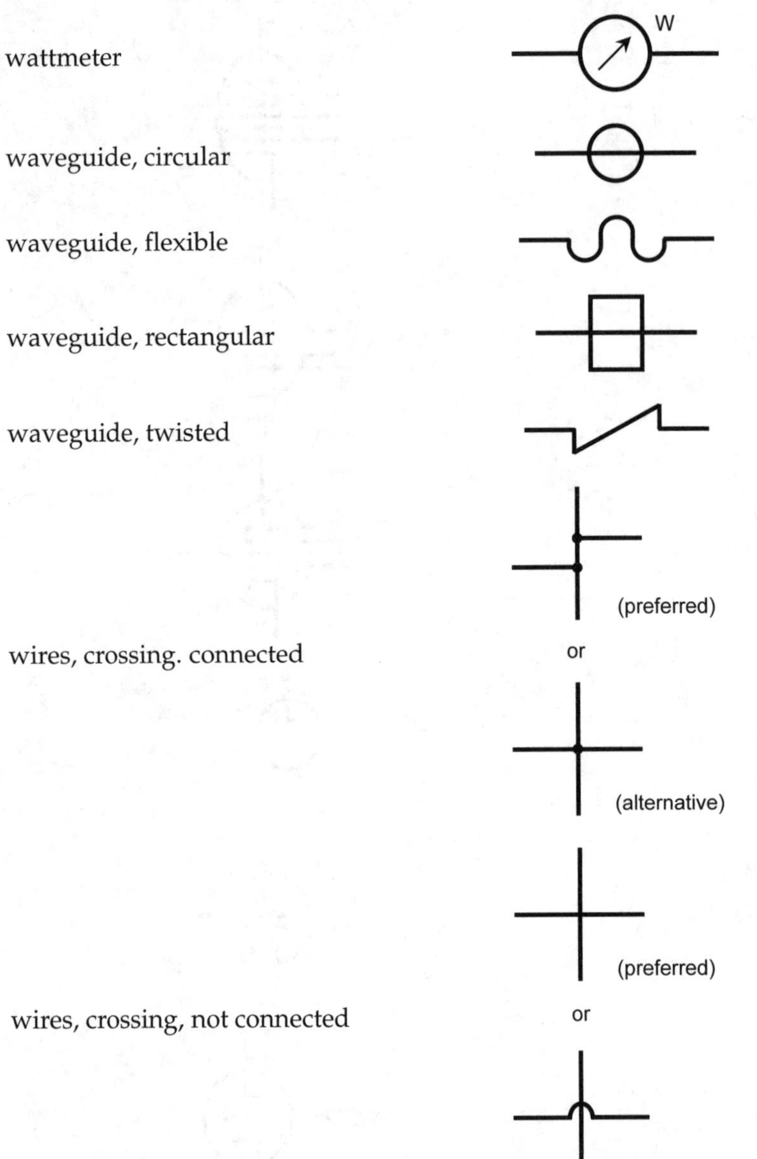
wattmeter
W
waveguide, circular
waveguide, flexible
waveguide, rectangular
waveguide, twisted
wires, crossing. connected
(preferred)
or
(alternative)
wires, crossing, not connected
(preferred)
or
(alternative)

APPENDIX B

Prefix Multipliers

Prefix Multipliers

Designator	Symbol	Decimal	Binary
yocto-	y	10^{-24}	2^{-80}
zepto-	z	10^{-21}	2^{-70}
atto-	a	10^{-18}	2^{-60}
femto-	f	10^{-15}	2^{-50}
pico-	p	10^{-12}	2^{-40}
nano-	n	10^{-9}	2^{-30}
micro-	µ or mm	10^{-6}	2^{-20}
milli-	m	10^{-3}	2^{-10}
centi-	c	10^{-2}	–
deci-	d	10^{-1}	–
(none)	–	10^{0}	2^{0}
deka-	da or D	10^{1}	–
hecto-	h	10^{2}	–
kilo-	K or k	10^{3}	2^{10}
mega-	M	10^{6}	2^{20}
giga-	G	10^{9}	2^{30}
tera-	T	10^{12}	2^{40}
peta-	P	10^{15}	2^{50}
exa-	E	10^{18}	2^{60}
zetta-	Z	10^{21}	2^{70}
yotta-	Y	10^{24}	2^{80}

APPENDIX C

Standard International Unit Conversions

Standard International Unit Conversions

To Convert:	To:	Multiply by:	Conversely, Multiply by:
meters (m)	Angstroms	10^{10}	10^{-10}
meters (m)	nanometers (nm)	10^{9}	10^{-9}
meters (m)	microns (μ)	10^{6}	10^{-6}
meters (m)	millimeters (mm)	10^{3}	10^{-3}
meters (m)	centimeters (cm)	10^{2}	10^{-2}
meters (m)	inches (in)	39.37	0.02540
meters (m)	feet (ft)	3.281	0.3048
meters (m)	yards (yd)	1.094	0.9144
meters (m)	kilometers (km)	10^{-3}	10^{3}
meters (m)	statute miles (mi)	6.214×10^{-4}	1.609×10^{3}
meters (m)	nautical miles	5.397×10^{-4}	1.853×10^{3}
meters (m)	light seconds	3.336×10^{-9}	2.998×10^{8}
meters (m)	astronomical units (AU)	6.685×10^{-12}	1.496×10^{11}
meters (m)	light years	1.057×10^{-16}	9.461×10^{15}
meters (m)	parsecs (pc)	3.241×10^{-17}	3.085×10^{16}
kilograms (kg)	atomic mass units (amu)	6.022×10^{26}	1.661×10^{-27}
kilograms (kg)	nanograms (ng)	10^{12}	10^{-12}
kilograms (kg)	micrograms (μg)	10^{9}	10^{-9}
kilograms (kg)	milligrams (mg)	10^{6}	10^{-6}
kilograms (kg)	grams (g)	10^{3}	10^{-3}
kilograms (kg)	ounces (oz)	35.28	0.02834

Standard International Unit Conversions (*Continued*)

To Convert:	To:	Multiply by:	Conversely, Multiply by:
kilograms (kg)	pounds (lb)	2.205	0.4535
kilograms (kg)	English tons	1.103×10^{-3}	907.0
seconds (s)	minutes (min)	0.01667	60.00
seconds (s)	hours (h)	2.778×10^{-4}	3.600×10^{3}
seconds (s)	days (dy)	1.157×10^{-5}	8.640×10^{4}
seconds (s)	years (yr)	3.169×10^{-8}	3.156×10^{7}
seconds (s)	centuries	3.169×10^{-10}	3.156×10^{9}
seconds (s)	millenia	3.169×10^{-11}	3.156×10^{10}
degrees Kelvin (°K)	degrees Celsius (°C)	Subtract 273	Add 273
degrees Kelvin (°K)	degrees Fahrenheit (°F)	Multiply by 1.80, then subtract 459	Multiply by 0.556, then add 255
degrees Kelvin (°K)	degrees Rankine (°R)	1.80	0.556
amperes (A)	carriers per second	6.24×10^{18}	1.60×10^{-19}
amperes (A)	statamperes (statA)	2.998×10^{9}	3.336×10^{-10}
amperes (A)	nanoamperes (nA)	10^{9}	10^{-9}
amperes (A)	microamperes (μA)	10^{6}	10^{-6}
amperes (A)	abamperes (abA)	0.10000	10.000
amperes (A)	milliamperes (mA)	10^{3}	10^{-3}
candela (cd)	microwatts per steradian (μW/sr)	1.464×10^{3}	6.831×10^{-4}
candela (cd)	milliwatts per steradian (mW/sr)	1.464	0.6831
candela (cd)	lumens per steradian (lum/sr)	identical; no conversion	identical; no conversion
candela (cd)	watts per steradian (W/sr)	1.464×10^{-3}	683.1
moles (mol)	coulombs (C)	9.65×10^{4}	1.04×10^{-5}

APPENDIX D

Electrical Unit Conversions

Electrical Unit Conversions

To Convert:	To:	Multiply by:	Conversely, Multiply by:
unit electric charges	coulombs (C)	1.60×10^{-19}	6.24×10^{18}
unit electric charges	abcoulombs (abC)	1.60×10^{-20}	6.24×10^{19}
unit electric charges	statcoulombs (statC)	4.80×10^{-10}	2.08×10^{9}
coulombs (C)	unit electric charges	6.24×10^{18}	1.60×10^{-19}
coulombs (C)	statcoulombs (statC)	2.998×10^{9}	3.336×10^{-10}
coulombs (C)	abcoulombs (abC)	0.1000	10.000
joules (J)	electronvolts (eV)	6.242×10^{18}	1.602×10^{-19}
joules (J)	ergs (erg)	10^{7}	10^{-7}
joules (J)	calories (cal)	0.2389	4.1859
joules (J)	British thermal units (Btu)	9.478×10^{-4}	1.055×10^{3}
joules (J)	watt-hours (Wh)	2.778×10^{-4}	3.600×10^{3}
joules (J)	kilowatt-hours (kWh)	2.778×10^{-7}	3.600×10^{6}
volts (V)	abvolts (abV)	10^{8}	10^{-8}
volts (V)	microvolts (μV)	10^{6}	10^{-6}
volts (V)	millivolts (mV)	10^{3}	10^{-3}
volts (V)	statvolts (statV)	3.336×10^{-3}	299.8
volts (V)	kilovolts (kV)	10^{-3}	10^{3}
volts (V)	megavolts (MV)	10^{-6}	10^{6}
ohms (Ω)	abohms (abΩ)	10^{9}	10^{-9}
ohms (Ω)	megohms (MΩ)	10^{-6}	10^{6}
ohms (Ω)	kilohms (kΩ)	10^{-3}	10^{3}

Electrical Unit Conversions (*Continued*)

To Convert:	To:	Multiply by:	Conversely, Multiply by:
ohms (Ω)	statohms (statΩ)	1.113×10^{-12}	8.988×10^{11}
siemens (S)	statsiemens (statS)	8.988×10^{11}	1.113×10^{-12}
siemens (S)	microsiemens (μS)	10^{6}	10^{-6}
siemens (S)	millisiemens (mS)	10^{3}	10^{-3}
siemens (S)	absiemens (abS)	10^{-9}	10^{9}
watts (W)	picowatts (pW)	10^{12}	10^{-12}
watts (W)	nanowatts (nW)	10^{9}	10^{-9}
watts (W)	microwatts (μW)	10^{6}	10^{-6}
watts (W)	milliwatts (mW)	10^{3}	10^{-3}
watts (W)	British thermal units per hour (Btu/hr)	3.412	0.2931
watts (W)	horsepower (hp)	1.341×10^{-3}	745.7
watts (W)	kilowatts (kW)	10^{-3}	10^{3}
watts (W)	megawatts (MW)	10^{-6}	10^{6}
watts (W)	gigawatts (GW)	10^{-9}	10^{9}
hertz (Hz)	degrees per second (deg/s)	360.0	0.002778
hertz (Hz)	radians per second (rad/s)	6.283	0.1592
hertz (Hz)	kilohertz (kHz)	10^{-3}	10^{3}
hertz (Hz)	megahertz (MHz)	10^{-6}	10^{6}
hertz (Hz)	gigahertz (GHz)	10^{-9}	10^{9}
hertz (Hz)	terahertz (THz)	10^{-12}	10^{12}
farads (F)	picofarads (pF)	10^{12}	10^{-12}
farads (F)	statfarads (statF)	8.898×10^{11}	1.113×10^{-12}
farads (F)	nanofarads (nF)	10^{9}	10^{-9}
farads (F)	microfarads (μF)	10^{6}	10^{-6}
farads (F)	abfarads (abF)	10^{-9}	10^{9}
henrys (H)	nanohenrys (nH)	10^{9}	10^{-9}
henrys (H)	abhenrys (abH)	10^{9}	10^{-9}
henrys (H)	microhenrys (μH)	10^{6}	10^{-6}
henrys (H)	millihenrys (mH)	10^{3}	10^{-3}
henrys (H)	stathenrys (statH)	1.113×10^{-12}	8.898×10^{11}
volts per meter (V/m)	picovolts per meter (pV/m)	10^{12}	10^{-12}
volts per meter (V/m)	nanovolts per meter (nV/m)	10^{9}	10^{-9}

Electrical Unit Conversions (*Continued*)

To Convert:	To:	Multiply by:	Conversely, Multiply by:
volts per meter (V/m)	microvolts per meter (μV/m)	10^{6}	10^{-6}
volts per meter (V/m)	millivolts per meter (mV/m)	10^{3}	10^{-3}
volts per meter (V/m)	volts per foot (v/ft)	3.281	0.3048
watts per square meter (W/m^2)	picowatts per square meter (pW/m^2)	10^{12}	10^{-12}
watts per square meter (W/m^2)	nanowatts per square meter (pW/m^2)	10^{9}	10^{-9}
watts per square meter (W/m^2)	microwatts per square meter ($\mu W/m^2$)	10^{6}	10^{-6}
watts per square meter (W/m^2)	milliwatts per square meter (mW/m^2)	10^{3}	10^{-3}
watts per square meter (W/m^2)	watts per square foot (W/ft^2)	0.09294	10.76
watts per square meter (W/m^2)	watts per square inch (W/in^2)	6.452×10^{-4}	1.550×10^{3}
watts per square meter (W/m^2)	watts per square centimeter (W/cm^2)	10^{-4}	10^{4}
watts per square meter (W/m^2)	watts per square millimeter (W/mm^2)	10^{-6}	10^{6}

APPENDIX E

Magnetic Unit Conversions

Magnetic Unit Conversions

To Convert:	**To:**	**Multiply by:**	**Conversely, Multiply by:**
webers (Wb)	maxwells (Mx)	10^{8}	10^{-8}
webers (Wb)	ampere-microhenrys (AμH)	10^{6}	10^{-6}
webers (Wb)	ampere-millihenrys (AmH)	10^{3}	10^{-3}
webers (Wb)	unit poles	1.257×10^{-7}	7.956×10^{6}
teslas (T)	maxwells per square meter (Mx/m^2)	10^{8}	10^{-8}
teslas (T)	gauss (G)	10^{4}	10^{-4}
teslas (T)	maxwells per square centimeter (Mx/cm^2)	10^{4}	10^{-4}
teslas (T)	maxwells per square millimeter (Mx/mm^2)	10^{2}	10^{-2}
teslas (T)	webers per square centimeter (W/cm^2)	10^{-4}	10^{4}
teslas (T)	webers per square millimeter (W/mm^2)	10^{-6}	10^{6}
oersteds (Oe)	microamperes per meter (μA/m)	7.956×10^{7}	1.257×10^{-8}
oersteds (Oe)	milliamperes per meter (mA/m)	7.956×10^{4}	1.257×10^{-5}
oersteds (Oe)	amperes per meter (A/m)	79.56	0.01257
ampere-turns (AT)	microampere-turns (μAT)	10^{6}	10^{-6}
ampere-turns (AT)	milliampere-turns (mAT)	10^{3}	10^{-3}
ampere-turns (AT)	gilberts (G)	1.256	0.7956

APPENDIX F

Miscellaneous Unit Conversions

Miscellaneous Unit Conversions

To Convert:	To:	Multiply By:	Conversely, Multiply By:
square meters (m^2)	square Angstroms	10^{20}	10^{-20}
square meters (m^2)	square nanometers (nm^2)	10^{18}	10^{-18}
square meters (m^2)	square microns (μ^2)	10^{12}	10^{-12}
square meters (m^2)	square millimeters (mm^2)	10^{6}	10^{-6}
square meters (m^2)	square centimeters (cm^2)	10^{4}	10^{-4}
square meters (m^2)	square inches (in^2)	1.550×10^{3}	6.452×10^{-4}
square meters (m^2)	square feet (ft^2)	10.76	0.09294
square meters (m^2)	acres	2.471×10^{-4}	4.047×10^{3}
square meters (m^2)	hectares	10^{-4}	10^{4}
square meters (m^2)	square kilometers (km^2)	10^{-6}	10^{6}
square meters (m^2)	square statute miles (mi^2)	3.863×10^{-7}	2.589×10^{6}
square meters (m^2)	square nautical miles	2.910×10^{-7}	3.434×10^{6}
square meters (m^2)	square light years	1.117×10^{-17}	8.951×10^{16}
square meters (m^2)	square parsecs (pc^2)	1.051×10^{-33}	9.517×10^{32}
cubic meters (m^3)	cubic Angstroms	10^{30}	10^{-30}
cubic meters (m^3)	cubic nanometers (nm^3)	10^{27}	10^{-27}
cubic meters (m^3)	cubic microns (μ^3)	10^{18}	10^{-18}
cubic meters (m^3)	cubic millimeters (mm^3)	10^{9}	10^{-9}
cubic meters (m^3)	cubic centimeters (cm^3)	10^{6}	10^{-6}
cubic meters (m^3)	milliliters (ml)	10^{6}	10^{-6}
cubic meters (m^3)	liters (l)	10^{3}	10^{-3}

Miscellaneous Unit Conversions (*Continued*)

To Convert:	To:	Multiply By:	Conversely, Multiply By:
cubic meters (m^3)	U.S. gallons (gal)	264.2	3.785×10^{-3}
cubic meters (m^3)	cubic inches (in^3)	6.102×10^4	1.639×10^{-5}
cubic meters (m^3)	cubic feet (ft^3)	35.32	0.02831
cubic meters (m^3)	cubic kilometers (km^3)	10^{-9}	10^9
cubic meters (m^3)	cubic statute miles (mi^3)	2.399×10^{-10}	4.166×10^9
cubic meters (m^3)	cubic nautical miles	1.572×10^{-10}	6.362×10^9
cubic meters (m^3)	cubic light seconds	3.711×10^{-26}	2.695×10^{25}
cubic meters (m^3)	cubic astronomical units (AU^3)	2.987×10^{-34}	3.348×10^{33}
cubic meters (m^3)	cubic light years	1.181×10^{-48}	8.469×10^{47}
cubic meters (m^3)	cubic parsecs (pc^3)	3.406×10^{-50}	2.936×10^{49}
radians (rad)	degrees (° or deg)	57.30	0.01745
meters per second (m/s)	inches per second (in/s)	39.37	0.02540
meters per second (m/s)	kilometers per hour (km/hr)	3.600	0.2778
meters per second (m/s)	feet per second (ft/s)	3.281	0.3048
meters per second (m/s)	statute miles per hour (mi/hr)	2.237	0.4470
meters per second (m/s)	knots (kt)	1.942	0.5149
meters per second (m/s)	kilometers per minute (km/min)	0.06000	16.67
meters per second (m/s)	kilometers per second (km/s)	10^{-3}	10^3
radians per second (rad/s)	degrees per second (°/s or deg/s)	57.30	0.01745
radians per second (rad/s)	revolutions per second (rev/s or rps)	0.1592	6.283
radians per second (rad/s)	revolutions per minute (rev/min or rpm)	2.653×10^{-3}	377.0
meters per second per second (m/s^2)	inches per second per second (in/s^2)	39.37	0.02540
meters per second per second (m/s^2)	feet per second per second (ft/s^2)	3.281	0.3048
radians per second per second (rad/s^2)	degrees per second per second ($°/s^2$ or deg/s^2)	57.30	0.01745

Miscellaneous Unit Conversions (*Continued*)

To Convert:	To:	Multiply By:	Conversely, Multiply By:
radians per second per second (rad/s^2)	revolutions per second per second (rev/s^2 or rps/s)	0.1592	6.283
radians per second per second (rad/s^2)	revolutions per minute per second (rev/min/s or rpm/s)	2.653×10^{-3}	377.0
newtons (N)	dynes	10^5	10^{-5}
newtons (N)	ounces (oz)	3.597	0.2780
newtons (N)	pounds (lb)	0.2248	4.448

APPENDIX G

American Wire Gauge (AWG) Diameters

American Wire Gauge (AWG) Diameters

AWG	Millimeters	Inches
1	7.35	0.289
2	6.54	0.257
3	5.83	0.230
4	5.19	0.204
5	4.62	0.182
6	4.12	0.163
7	3.67	0.144
8	3.26	0.128
9	2.91	0.115
10	2.59	0.102
11	2.31	0.0909
12	2.05	0.0807
13	1.83	0.0720
14	1.63	0.0642
15	1.45	0.0571
16	1.29	0.0508
17	1.15	0.0453
18	1.02	0.0402
19	0.912	0.0359
20	0.812	0.0320
21	0.723	0.0285
22	0.644	0.0254
23	0.573	0.0226

American Wire Gauge (AWG) Diameters (*Continued*)

AWG	Millimeters	Inches
24	0.511	0.0201
25	0.455	0.0179
26	0.405	0.0159
27	0.361	0.0142
28	0.321	0.0126
29	0.286	0.0113
30	0.255	0.0100
31	0.227	0.00894
32	0.202	0.00795
33	0.180	0.00709
34	0.160	0.00630
35	0.143	0.00563
36	0.127	0.00500
37	0.113	0.00445
38	0.101	0.00398
39	0.090	0.00354
40	0.080	0.00315

APPENDIX H

British Standard Wire Gauge (NBS SWG) Diameters

British Standard Wire Gauge (NBS SWG) Diameters

NBS SWG	Millimeters	Inches
1	7.62	0.300
2	7.01	0.276
3	6.40	0.252
4	5.89	0.232
5	5.38	0.212
6	4.88	0.192
7	4.47	0.176
8	4.06	0.160
9	3.66	0.144
10	3.25	0.128
11	2.95	0.116
12	2.64	0.104
13	2.34	0.092
14	2.03	0.080
15	1.83	0.072
16	1.63	0.064
17	1.42	0.056
18	1.22	0.048
19	1.02	0.040
20	0.91	0.036

British Standard Wire Gauge (NBS SWG) Diameters (*Continued*)

AWG	Millimeters	Inches
21	0.81	0.032
22	0.71	0.028
23	0.61	0.024
24	0.56	0.022
25	0.51	0.020
26	0.46	0.018
27	0.42	0.0164
28	0.38	0.0148
29	0.345	0.0136
30	0.315	0.0124
31	0.295	0.0116
32	0.274	0.0108
33	0.254	0.0100
34	0.234	0.0092
35	0.213	0.0084
36	0.193	0.0076
37	0.173	0.0068
38	0.152	0.0060
39	0.132	0.0052
40	0.122	0.0048

APPENDIX I

Birmingham Wire Gauge (BWG) Diameters

Birmingham Wire Gauge (BWG) Diameters

BWG	Millimeters	Inches
1	7.62	0.300
2	7.21	0.284
3	6.58	0.259
4	6.05	0.238
5	5.59	0.220
6	5.16	0.203
7	4.57	0.180
8	4.19	0.165
9	3.76	0.148
10	3.40	0.134
11	3.05	0.120
12	2.77	0.109
13	2.41	0.095
14	2.11	0.083
15	1.83	0.072
16	1.65	0.064
17	1.47	0.058
18	1.25	0.049
19	1.07	0.042
20	0.889	0.035

APPENDIX J

Maximum Safe DC Carrying Capacity for Bare Copper Wire in Open Air

Maximum Safe DC Carrying Capacity for Bare Copper Wire in Open Air

Wire Size, AWG	Current, A
8	73
9	63
10	55
11	47
12	41
13	36
14	31
15	26
16	22
17	18
18	15
19	13
20	11

APPENDIX K
Common Solder Alloys

Common Solder Alloys

Solder Type	Melting Point	Common Uses
Tin/lead 50/50 rosin core	430° F 220° C	Electronics
Tin/lead 60/40 rosin core, low heat	370° F 190° C	Electronics
Tin/lead 63/37 rosin core, low heat	360° F 180° C	Electronics
Silver high heat, high current	600° F 320° C	Electronics
Tin/lead 50/50 acid core	430° F 220° C	Sheet-metal bonding

APPENDIX L

Radio Frequency Bands

Radio Frequency Bands

Standard Designation	Frequency Range	Wavelength Range
Very Low (VLF)	3 kHz – 30 kHz	100 km – 10 km
Low (LF)	30 kHz – 300 kHz	10 km – 1 km
Medium (MF)	300 kHz – 3 MHz	1 km – 100 m
High (HF)	3 MHz – 30 MHz	100 m – 10 m
Very High (VHF)	30 MHz – 300 MHz	10 m – 1 m
Ultra High (UHF)	300 MHz – 3 GHz	1 m – 100 mm
Super High (SHF)	3 GHz – 30 GHz	100 mm – 10 mm
Extremely High (EHF)	30 GHz – 300 GHz	10 mm – 1 mm

Suggested Additional Reading

American Radio Relay League, Inc. *The ARRL Handbook for Radio Communications*. Newington, CT: ARRL, revised annually. In addition to the *Handbook*, a classic in its field, I recommend all ARRL publications on topics that interest you! You can buy their books and training materials through their website at www.arrl.org.

Frenzel, Louis E., Jr., *Electronics Explained*. Burlington, MA: Newnes/Elsevier, 2010.

Geier, Michael, *How to Diagnose and Fix Everything Electronic*. New York: McGraw-Hill, 2011.

Gerrish, Howard, *Electricity and Electronics*. Tinley Park, IL: Goodheart-Wilcox Co., 2008.

Gibilisco, Stan, *Beginner's Guide to Reading Schematics*, 3rd ed. New York: McGraw-Hill, 2014.

Gibilisco, Stan, *Electricity Experiments You Can Do at Home*. New York: McGraw-Hill, 2010.

Gibilisco, Stan, *Electronics Demystified*, 2nd ed. New York: McGraw-Hill, 2011.

Gibilisco, Stan, *Ham and Shortwave Radio for the Electronics Hobbyist*. New York: McGraw-Hill, 2014.

Gibilisco, Stan, *Teach Yourself Electricity and Electronics*, 5th ed. New York: McGraw-Hill, 2011.

Horn, Delton, *Basic Electronics Theory with Experiments and Projects*, 4th ed. New York: McGraw-Hill, 1994.

Horn, Delton, *How to Test Almost Everything Electronic*, 3rd ed. New York: McGraw-Hill, 1993.

Kybett, Harry, *All New Electronics Self-Teaching Guide*, 3rd ed. Hoboken, NJ: Wiley Publishing, 2008.

Mims, Forrest M., *Getting Started in Electronics*. Niles, IL: Master Publishing, 2003.

Index

B

C

D

J

K

L

P

Q

R

T